humankind – from
humble farmers in the
fields and toiling workers
in the cities to teachers,
people of independent
means, those who have reached the
pinnacle of fame or fortune, even the
most frivolous of society women – if
they knew what profound inner
pleasure awaits those who gaze at the
heavens, then France, nay, the whole
of Europe, would be covered
with telescopes instead
of bayonets, thereby
promoting universal
happiness and peace. **"**

Camille Flammarion, 1880
French astronomer

CONTENTS

NEWTON
UNDERSTANDING THE COSMOS

Jean-Pierre Maury

THAMES AND HUDSON

In June 1665, as the Great Plague ran its deadly course, Cambridge University closed down and sent its students and professors home. Among them was Isaac Newton, a young man who had just received his bachelor's degree. He set off for the peace and quiet of his rural English birthplace, where he would spend a year. This period was so rich in discovery that future historians would refer to it as the *annus mirabilis*, the 'miraculous year'.

CHAPTER 1
ISAAC NEWTON'S HOLIDAY

Newton's year-long enforced holiday – without a doubt the most fruitful in the history of science – was spent in this house in Lincolnshire.

Isaac Newton was born on Christmas Day, 1642, the year Galileo died. His father had died a few months earlier. Three years later, his mother remarried and moved to a neighbouring village, leaving Isaac in the care of his grandmother at Woolsthorpe, an estate near Grantham, Lincolnshire, that had been in the family for two hundred years.

When Newton was fourteen, his mother, widowed a second time, returned to Woolsthorpe with the three children of her second marriage. Soon afterwards she brought Isaac home from school to learn to manage the estate, which did not appeal to him in the least. He divided his time between reading and ingenious tinkering – he built doll's-houses for his little sisters, a model windmill, a water clock that continued to run for years – so his mother decided to send him back to school, and at eighteen he was admitted to Trinity College at Cambridge University. He had just completed his studies when the plague forced him to go back home.

In the country Newton resumes the experiments with light which he began at Cambridge

Since Newton had started keeping track of his readings, experiments and ideas in notebooks by 1664, we know that at this time he was already mulling over Galileo's *Dialogue on the Two Chief World-Systems*, René Descartes' *Geometry* and Johannes Kepler's work, particularly his research with light.

The fact that a ray of sunlight passing through a glass prism produces colours had been known for quite some time.

The Great Plague of 1665 claimed more than seventy thousand victims in London alone. This contemporary broadsheet shows Londoners fleeing their city by all available escape routes, burial processions and carts filled with corpses to be buried.

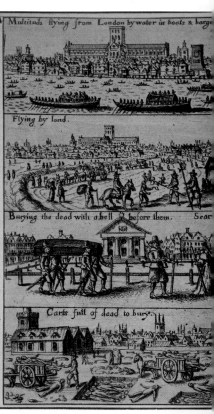

Multituds flying from London by water in boats & barge

Flying by land.

Burying the dead with a bell before them.   Sear

Carts full of dead to bury.

Such an experiment with a prism had appeared in Giambattista della Porta's *De refractione,* which was published in Naples in 1558. But the phenomenon was explained in terms of the old Aristotelian concept that light is white and that colours arise through its gradual modification. Red and yellow light, the colours of fire, were said to be modified the least. Then came green, blue and violet, each 'mingled' with increasing amounts

Ioannis Keppleri

HARMONICES
MVNDI

LIBRI V. QVORVM

Primus GEOMETRICVS, De Figurarum Regularium, quæ Proportiones Harmonicas conftituunt, ortu & demonftrationibus.

Secundus ARCHITECTONICVS, feu ex GEOMETRIA FIGVRATA, De Figurarum Regularium Congruentia in plano vel folido:

Tertius propriè HARMONICVS, De Proportionum Harmonicarum ortu ex Figuris; deque Naturâ & Differentiis rerum ad cantum pertinentium, contra Veteres:

Quartus METAPHYSICVS, PSYCHOLOGICVS & ASTROLOGICVS, De Harmoniarum mentali Effentia earumque generibus in Mundo; præfertim de Harmonia radiorum, ex corporibus cœleftibus in Terram defcendentibus, eiufque effectu in Natura feu Anima fublunari & Humana:

Quintus ASTRONOMICVS & METAPHYSICVS, De Harmoniis abfolutiffimis motuum cœleftium, ortuque Eccentricitatum ex proportionibus Harmonicis.

Here is some of young Newton's bedside reading: Kepler's *Harmonices Mundi* (1619) and, above all, Galileo's *Dialogue on the Two Chief World-Systems* (1632). In the first, buried in a mass of geometric, aesthetic and metaphysical speculation, Newton found the three laws of planetary motion, laws that he would be the first to demonstrate conclusively. The second, which had earned Galileo the condemnation of the Church, was later acclaimed as the manifesto of modern astronomy. Relieved at last of their 'heavenly perfection', celestial bodies were no longer placed above scientific reasoning.

of 'darkness'. If a beam of white light passing through a prism turned red near the apex and blue near the base, it was said to be because of the varying thickness of the glass through which the light passed: the lower end of the beam passed through more glass and, thus modified to a greater degree, turned blue. By Newton's time, of course, water droplets were known to produce rainbows by a process that was assumed to be the same.

Newton gave all this a great deal of thought. At first he investigated early theories and simply tried to expand upon them. Perhaps this 'modification' of light could be explained as a purely mechanical process (for example, perhaps glass slows light down), and so he attempted to grind lenses in shapes that would inhibit the colour change. He distinguished rays of different colours, which had never been done before.

Newton discovers the cause: 'white' light is actually a mixture of different colours of light

And a prism refracts each one by a different amount. He described his discovery in a letter a few years later. 'In the beginning of the Year 1666... I procured me a Triangular glass-Prisme, to try therewith the celebrated

Finding rainbows intriguing, people have long ascribed supernatural properties to them. Scientists started investigating them during the Middle Ages, but Newton was the first to explain them correctly. In the image opposite, the water droplets dispersing the colours are being produced by a waterfall.

A 1748 depiction of a rainbow seen in the Peruvian Andes (below).

Phaenomena of Colours. And in order thereto having darkened my chamber, and made a small hole in my window-shuts, to let in a convenient quantity of the Sun's light, I placed my Prisme at his entrance, that it might be thereby refracted to the opposite wall. It was at first a very pleasing divertisement, to view the vivid and intense Colours produced thereby; but after a while [I applied] myself to consider them more circumspectly.'

Newton was twenty-five when he discovered that the 'white' light of the sun is, in fact, a mixture of light of different colours.

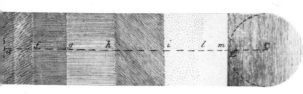

The apparatus below is far more elaborate than the one Newton used in his room at Woolsthorpe: a simple hole in a window shutter, a prism and a wall on which to project the spectrum.

The first thing he noticed was that the spot of light was not only multicoloured but oblong, and the prism bent the 'blue end' more than it did the red. Could this have been caused by a flaw in the prism? How could he tell? Newton placed a second prism next to the first, only facing the other way, so that they would compensate for each other's powers but not their flaws. The resulting spot of light was white and circular, indicating that there were no defects in the prism.

Gradually, Newton was led to conduct what he referred to as his 'crucial experiment'. He isolated the blue part of the light coming in through a hole in the shutter and then through a prism and passed it through a second prism. It was, of course, refracted. But it was not spread out, and its colour did not change!

This time, Newton was sure: 'white' sunlight is, in fact, a mixture of many different colours of light, each of which refracts to a different degree when passed through a prism. He then conducted additional experiments demonstrating, in particular, several ways in which mixtures of spectral colours could yield so-called white light.

Oddly enough, Newton keeps silent about his extraordinary discovery

Undoubtedly there were several reasons for his reticence. To begin with, Newton knew that a

discovery this
revolutionary would arouse
controversy in the scientific
community. He did publish his findings, but not until
1672, after he had become a professor and his
invention of the reflecting telescope had earned him
the acclaim of his colleagues.

This reluctance held true throughout Newton's life.
He always made his discoveries known only under
duress, after constant prodding. No doubt he was
conscientious about building up an arsenal of
experiments and proofs. Mainly it was because he was
reclusive and diffident by nature. He shrank from
controversy and the hurly-burly of polemics.

And if he kept the discovery of the composite nature
of light under wraps for five years, an even more
momentous discovery – the law of universal gravitation,
surely the greatest of all the miracles from the *annus
mirabilis* of 1665–6 – remained undisclosed for
twenty years.

Probably no apple (except Eve's, of course) has inspired more artwork than Newton's. Some illustrators showed it hitting the English scientist on the head; in other cases, such as this 19th-century engraving (right), we see him pondering an already-fallen fruit. In any event, the problem involved not so much observing the apple as comparing it to the moon.

Newton contemplates the apple and the moon and discovers the mainspring of the universe

As is often the case with flashes of genius that spawn
earthshaking change, the process by which the concept
of universal gravitation came to be conceived has come
down to us in the form of an anecdote that may well be
apocryphal. Then again, you never can tell.

One mild autumn evening, a pensive Newton was
gazing at the moon under an apple tree at Woolsthorpe.

Suddenly, an apple fell, because everything deprived of support falls to the ground.

But what about the moon, he wondered. It is not held up by anything, so why doesn't it fall, too? An explanation flashed into Newton's mind: it *does* fall!

If it were not falling towards the earth, it would keep moving in a straight line and fly off into space. The combination of its forward movement and the pull exerted by the earth causes the moon to revolve around the earth, just as the earth and the other planets revolve

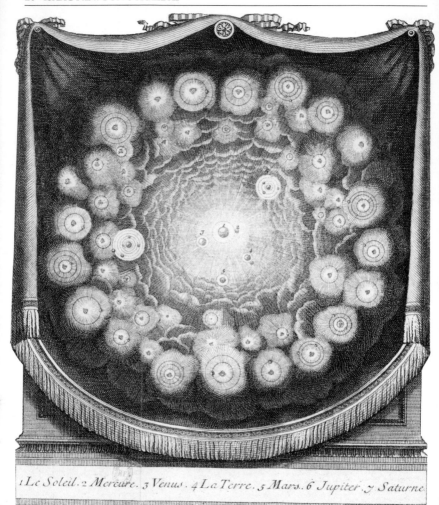

1 Le Soleil. 2 Mercure. 3 Venus. 4 La Terre. 5 Mars. 6 Jupiter. 7 Saturne.

around the sun. The satellites of Jupiter revolve around
their planet. Titan, which the Dutch astronomer
Christiaan Huygens (1629–95) had recently
discovered, revolves around Saturn. Newton wondered
if the immense workings of the solar system and the
apple falling could have a common explanation.

Newton's contemporaries are unable to account for the origin of comets

The year was 1665. A little over one hundred and twenty years earlier, Nicholas Copernicus had published his theory that the planets revolve around the sun. Fifty years earlier, Johannes Kepler had formulated laws describing their motion.

Thirty years earlier, Galileo had been condemned by the Church for making all that palpable, for turning his telescope into an instrument of discovery. He had torn down the barrier that had stood between the earth and the heavens for some two thousand years.

Since the time of Plato and Aristotle, astronomy and physics had been kept apart. It was forbidden to seek the natural causes of the motion of heavenly bodies, motion which, like the bodies themselves, was considered 'perfect'.

By showing that there were craters on the moon and blemishes on the sun, Galileo shattered this 'perfection'. The moon was very like the earth, hence, no more perfect than it was. Why, then, should its motion be perfect and less subject to the forces that cause everyday objects to move? The idea that the same natural law might govern the moon and the apple was sacrilege in Galileo's time. Although Galileo suffered for his belief, thanks to him, the existence of that

The two most peculiar aspects of this 18th-century engraving (opposite) are the profusion of planetary systems and the thick clouds – cosmic darkness materialized – which the light of the sun has caused merely to recede. More than two thousand years after Plato and Aristotle had dissipated it, the 'misty darkness' of antiquity lived on in the imagination of artists.

Galileo showed the senators of Venice that his spyglass could be used to see distant ships and monuments. Before long, however, he was training his new instrument on the heavens.

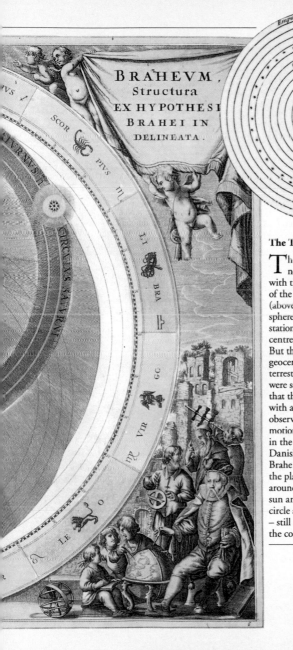

Empiree où Sejour des Bienheureux
le Premier Mobile
Premier Cristallin
Second Cristallin
Ciel des Etoiles fixées
Ciel de Saturne
Ciel de Jupiter
Ciel de Mars
Ciel du Soleil
Ciel de Venus
Ciel de Mercure
Ciel de la Lune
Terre
l'Air
le Feu

BRAHEVM,
Structura
EX HYPOTHESI
BRAHEI IN
DELINEATA.

The Tychonic system

The Renaissance was no longer satisfied with the perfect order of the Ptolemaic cosmos (above), with its nested spheres around a stationary earth at the centre of the universe. But the precepts of geocentricity and terrestrial immobility were so deeply entrenched that they endured, even with advances in the observation of planetary motion. For example, in the 16th century the Danish astronomer Tycho Brahe maintained that the planets revolve around the sun, while the sun annually describes a circle around the earth – still at the centre of the cosmos.

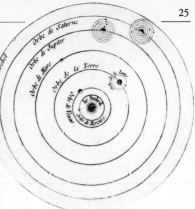

Echelle du diametre de la Terre
Contenant 2865 lieues de 2283 toises
Pour servir à mesurer les diametres
des autres Planetes

Diametre de 1180 lieues
Mercure

Diam. de 2305 lieues
Venus

Diametre de 1921 lieues
Mars

La Diametre de Jupiter est de 22664 lieues
Jupiter

La globe de Saturne est de 18235 lieues
Saturne

The Copernican system

With the Copernican system (published 1543), everything became simpler again, except that the earth was seen as neither immobile nor central: it revolved around the sun, along with the other planets. Improved instruments soon gave people an idea of what the planets looked like; but before their sizes could be known, their distances – that is, the dimensions of the solar system – had to be measured. And before one absolute measurement could be used to deduce all the others, it was necessary to determine (in 1672) the relative distance between the sun and the earth, the sun and Mars, and so on. Above: Uranus, still called Herschel (after its discoverer, William Herschel), appears in this engraving.

natural law became self-evident a few years later.

René Descartes (1596–1650), for one, tried to come up with a natural explanation for the motion of heavenly bodies. Since, like all his contemporaries, he was unwilling to accept the concept of action at a distance, he maintained that the empty space between heavenly bodies was filled with vortices of invisible matter, rotating whorls capable of carrying all the planets and satellites along in the same direction.

However – and Newton was fully aware of this – some comets were known to be 'wayward' and revolved in the opposite direction. Either they defied the vortices, or else the vortices did not exist.

Would it therefore be necessary to acknowledge that action at a distance might exist, that the earth might attract the moon, hundreds of thousands of miles away?

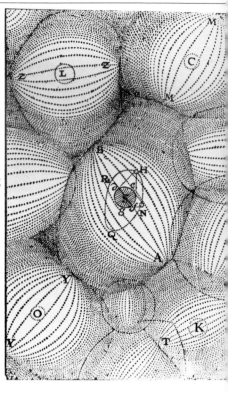

Newton searches for a law: how does that attraction vary with distance?

Distance from exactly what? Another 'miraculous' idea: for the apple and the moon alike, it was the distance to the *centre* of the earth that mattered. The apple was approximately 6400 kilometres away from the centre, the moon approximately 380,000. All Newton had to do was calculate the distance each 'fell' in one second. He formulated the law of universal gravitation: the attraction between bodies is inversely proportional to the square of the distance between their centres.

Was this law confirmed by the gravitational pull of the sun on the various planets? Yes, roughly speaking,

Although the Church posed less of a threat to Descartes after he settled in Holland in 1628, he did not publish his 'system of the world' (opposite above) for fear of suffering the same fate as Galileo, who was condemned by the Church in 1633. Descartes maintained that the spaces between heavenly bodies were filled with vortices of invisible matter (above).

DESCARTES COMPOSANT SON SYSTÈME DU MONDE.

René Descartes, *né à la Haye en Touraine en 1596,* | *la gloire de la France. Christine, Reine de Suède, fut plus* ?

if one could assume that their orbits are circular. Kepler, however, had already shown that the orbits are elliptical.

To study this further Newton would not only have to create a new branch of mathematics, but also further develop Galileo's theories of force and motion. He had his work cut out for him – for years to come! And, as we know, he was in no hurry to let anyone know about his discoveries, much less publish them. It would be nearly twenty years before the world learned of the law of universal gravitation.

During those twenty years, the face of European astronomy was to change irreversibly, as new and exciting discoveries were made.

The moon's diameter (below) is roughly one-quarter the earth's; they are c. 30 terrestrial diameters apart.

" On Tuesday, 21 June 1667, the day of the solstice, Messieurs Auzout, Frenicle, Picard, Buot and Richer went to the Uranoscope, or Observatory, in the morning to mark a meridian line on a stone.... " Thus began the construction of the Paris Observatory. In less than ten years, scientists there were to measure the circumference of the earth, the distance from the earth to the sun and the speed of light.

CHAPTER 2
THE BIRTH OF MODERN ASTRONOMY

In 1667 many European scientists were active in Paris. With the construction of the Paris Observatory, the city became one of the preeminent astronomical centres of its day.

When Jean Baptiste Colbert officially founded the French Academy of Sciences in 1666, scientists in France and other countries had already begun meeting on a regular basis, usually in someone's house, to discuss their work and discoveries.

In Paris, these sessions were held in the home of a curious figure by the name of Melchisédec Thévenot, a scientific jack-of-all-trades, indefatigable collector and noted traveller. It was under his roof that Marin Mersenne, who corresponded with scientists throughout Europe, introduced Thomas Hobbes, the English philosopher, to Descartes. This circle also included Pierre Gassendi and Blaise Pascal, both philosophers and scientists.

Across Europe, similar groups already enjoyed royal patronage or official status. As early as 1560 the

In the scene below, Jean Baptiste Colbert presents the members of the French Academy of Sciences to King Louis XIV on the occasion of the founding of the Paris Observatory. Although this particular event did not actually occur – Louis XIV did not visit the Observatory until 1682 – the Academy of Sciences and the Paris Observatory were established at Louis' behest and on Colbert's initiative.

Accademia Secretorum was founded in Naples, followed by the Accademia dei Lincei in Rome (1603) and the Accademia del Cimento in Florence (1657). The Holy Roman Emperor Leopold I was a patron of a learned society in Bavaria, and England's Royal Society, which had been meeting informally since 1645, was organized in 1662.

Of the seven founding members of the French Academy of Sciences, no fewer than four were astronomers: Adrien Auzout, Jean Picard, Gilles de Roberval and Christiaan Huygens of Holland, who had recently come to Paris at the invitation of Louis XIV. Huygens not only discovered Saturn's rings, but was the first to use pendulum clocks, a development that was to revolutionize astronomy.

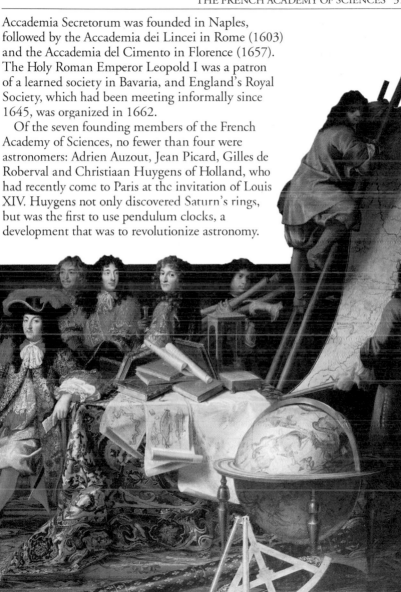

The accuracy of astronomical measurement suddenly increases

In his early experiments, the only way Galileo could count a succession of equal time intervals was to use the beats of his own pulse as a clock. No doubt this was how he studied pendulum oscillations (while watching a swinging church lamp during a tedious mass, as legend has it). At any rate, he noticed that, for small oscillations, the time of one complete pendulum swing (period) is independent of oscillation width (amplitude).

This observation gave Huygens the idea of using a pendulum to control the movement of a clock. Until then, mechanical clocks were regulated by the tension of an elastic strip – a highly imprecise way of measuring time. They were even less reliable than the kind of water clock young Isaac Newton had built at Woolsthorpe.

However, with the advent of the self-regulating pendulum, which uses an escapement mechanism to maintain a uniform swinging motion, everything changed. By around 1665, mechanical clocks were accurate to within one second every twenty-four hours – a thousandfold improvement.

Auzout's micrometer was a device with a screw which, when turned, moved a wire back and forth in front of a telescopic eyepiece so that it could be positioned directly over the image of a star. By registering on a graduated drum the angle through which the screw was turned, the displacement of the wire could be measured in increments as small as a

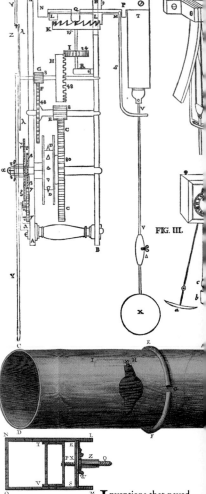

FIG. III.

Inventions that paved the way for precise astronomical measurement: Christiaan Huygens' pendulum clock (top) and Adrien Auzout's sliding-frame screw micrometer.

hundredth of a millimetre. Measurements of position were one hundred times more accurate than before!

In 1665 Auzout advises Louis XIV to build an observatory

'The glory of Your Majesty and the reputation of France are at stake, which gives us reason to hope that Your Majesty will provide for a site where in future all manner of celestial observations may be conducted, and that Your Majesty will equip it with all the necessary instruments.'

In 1667 Colbert purchased on behalf of the Crown 'a parcel of land with a windmill on it ... outside Saint James's [blind] gate, at a place known as Grand Regard.' On the day of the summer solstice, the astronomers of the Academy of Sciences gathered at the site, ceremoniously decided which way the future building should face, and marked on the ground a line that would come to be known as the Paris meridian.

Though the building was not finished until 1672, a list of projects had been drawn up since 1667. The first item was to make a precise measurement of the earth's dimensions, and astronomical research began straightaway.

The astronomers of the Paris Observatory do not reach for the stars until they have measured the earth

Since the earth was assumed to be spherical, scientists knew that its size could be determined by measuring the length of one degree of meridian, that is, the distance between two points along the same meridian separated by one degree of latitude. To compute the circumference of the globe, one would simply multiply that figure by 360.

Huygens (1629–95) was one of the leading physicists of the 17th century. Famous at an early age for his pendulum clock and discovery of Saturn's rings and first satellite, Huygens also published the first paper on the theory of probability, made substantial contributions to the mathematics and dynamics of curves, and laid the groundwork for the wave theory of light.

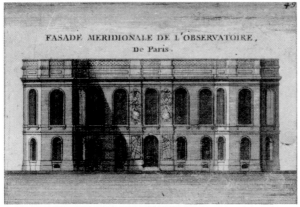

FASADE MERIDIONALE DE L'OBSERVATOIRE,
De Paris.

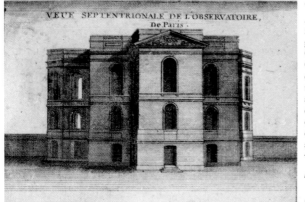

VEUE SEPTENTRIONALE DE L'OBSERVATOIRE,
De Paris.

Louis XIV ordered the Marly Tower (above right) to be moved to the gardens of the Paris Observatory to house Cassini's lenses. Instead of using tubes for his telescopes, Cassini moved about the garden, eyepiece in hand, observing through an objective several metres away.

Two wings and two cupolas were added to the Observatory in about 1835, but the main building looks much as it did in the 17th century (above left).

The path of a total solar eclipse is never very wide, so the necessary instruments must be brought to a different site (and makeshift shelters built for them) every time (right).

The distance involved, approximately equal to 110 kilometres, would have to be measured with the utmost precision. The method chosen was triangulation, which Willebrord Snell (Snellius), a Dutch mathematician and Descartes' competitor in the field of optics, had previously applied in 1615. Triangulation is based on the principle that once the angles and one side of a triangle are known, it is easy to calculate the length of the other two sides. Thus, by dividing the target area – in this case, the section of meridian – into a system of many triangles, one could ultimately determine its length.

It takes Picard two years to measure the meridian arc between Paris and Amiens

The Abbé Jean Picard once served as Gassendi's assistant (together they observed the solar eclipse of 21 August 1645) and helped to perfect Auzout's micrometer. A founding member of the Academy of Sciences, Picard supervised the construction of the first instruments installed at the Paris Observatory. (His directions for their use and regulation, dating from 1671, are still in use today.) In the meantime, he applied himself to measuring the earth. Fortunately, he did so near Paris, which enabled him to

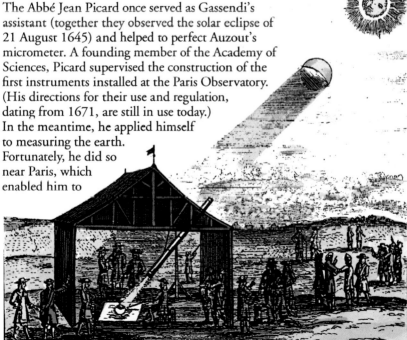

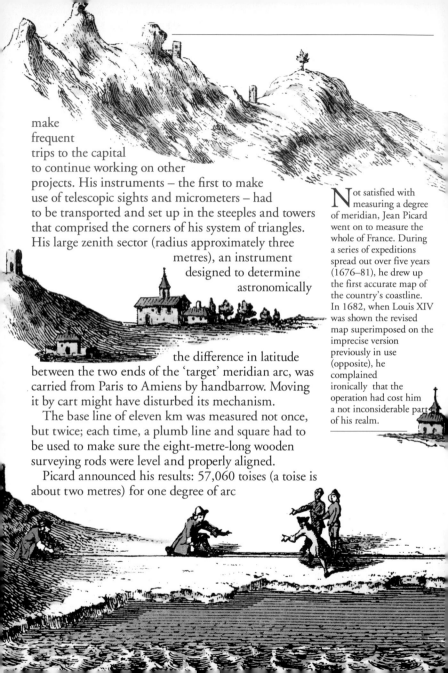

make
frequent
trips to the capital
to continue working on other
projects. His instruments – the first to make
use of telescopic sights and micrometers – had
to be transported and set up in the steeples and towers
that comprised the corners of his system of triangles.
His large zenith sector (radius approximately three
metres), an instrument
designed to determine
astronomically
the difference in latitude
between the two ends of the 'target' meridian arc, was
carried from Paris to Amiens by handbarrow. Moving
it by cart might have disturbed its mechanism.

The base line of eleven km was measured not once,
but twice; each time, a plumb line and square had to
be used to make sure the eight-metre-long wooden
surveying rods were level and properly aligned.

Picard announced his results: 57,060 toises (a toise is
about two metres) for one degree of arc

Not satisfied with
measuring a degree
of meridian, Jean Picard
went on to measure the
whole of France. During
a series of expeditions
spread out over five years
(1676–81), he drew up
the first accurate map of
the country's coastline.
In 1682, when Louis XIV
was shown the revised
map superimposed on the
imprecise version
previously in use
(opposite), he
complained
ironically that the
operation had cost him
a not inconsiderable part
of his realm.

– an important figure for Newton, as we shall see. This came to within about 0.1 per cent of the correct value.

The Paris Observatory needs an experienced director: Colbert sends for Gian Domenico Cassini

In 1669, the year Colbert summoned him to Paris, Cassini had been professor of astronomy at the University of Bologna for fifteen years. He was conversant with all areas of astronomy. His tables of the movements of Jupiter's satellites, published in 1668, were the best available and of great importance because at the time they were the only accurate means of determining longitude.

In 1671, with the building still unfinished, Cassini set up shop in the Paris Observatory, installed first-rate equipment – Italian objectives (lenses) were still unrivalled – and promptly made a name for the institution by discovering Iapetus, one of Saturn's satellites, that very same year. (He discovered Rhea in 1672 and Tethys and Dione twelve years later, in

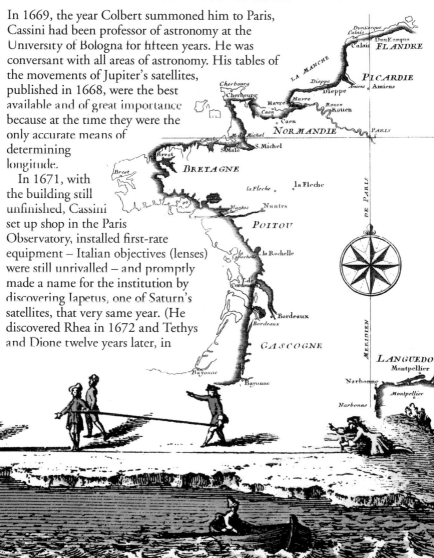

The Cassini dynasty directed the Paris Observatory from 1669, the year Gian Domenico (Cassini I, left) came from Bologna at Colbert's invitation, to 1793, the year Cassini IV resigned. In terms of astronomical research, the combined work of the last three cannot compare with the first Cassini's achievements. In addition to his many discoveries pertaining to Saturn (four of its satellites, the division of its rings) and measurement of the distance to the sun with Richer, he produced a large map of the moon, attempted to calculate the rotational period of Venus and continually refined his tables of the satellites of Jupiter.

1684.) He also discovered that Saturn's rings were divided in two and, as we shall soon see, collaborated with Jean Richer in the very first experiment to determine the distance between the earth and the sun.

Picard travels to a Danish island to determine the exact location of Tycho Brahe's observatory, Uraniborg, now in ruins

In 1669 Picard stated the astronomical goals that made this expedition necessary: 'So that the experiments to be conducted here may be compared with those of Tycho Brahe and that the Uraniborg meridian may be replaced by that of Paris, it is necessary to determine

Cassini also supervised the construction and equipping of the Paris Observatory, served as its scientific and administrative director, and trained an entire generation of astronomers. Clearly, the day Colbert invited him to move from Bologna to Paris proved a lucky one indeed.

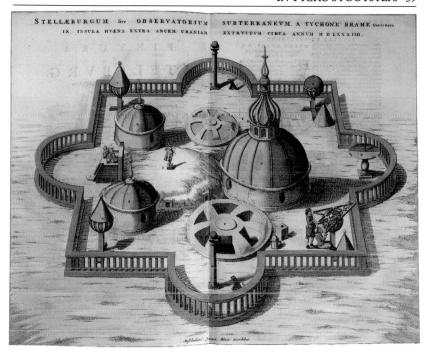

STELLÆBURGUM sive OBSERVATORIUM SUBTERRANEVM A TYCHONE BRAHE Nobili Dano IN INSULA HVÆNA EXTRA ARCEM URANIAM EXTRVCTVM CIRCA ANNVM M D LXXXIIII.

precisely the difference in longitude between these two meridians, and to that end we need to obtain information about the satellites of Jupiter at these two sites respectively. Furthermore, we would do well to remeasure the altitude of the [celestial] pole from the spot at which Uraniborg once stood, both to compare our instruments with the ones Tycho used and to test the reliability of his observations.'

Taking into consideration the importance of Tycho's observations – the basis for Kepler's laws – it was now an intriguing prospect to be able to compare them with those that would be made in Paris, that is, to correlate positional data recorded a century apart.

For this operation, the eclipses of Jupiter's satellites acted like a time signal on the radio. Simultaneously visible in Paris and Denmark, the eclipses allowed observers to compare local times at the two sites.

In 1572 Tycho Brahe, then twenty-six, became famous by discovering the first nova of modern times. In 1577 the king of Denmark granted him title to the island of Hven, where he built his observatory, Uraniborg ('heavenly castle', above). But the king died, and Tycho fell out with his successors and was forced to settle in Prague. His observatory deteriorated into a pile of rubble.

EFFIGIES TYCHONIS BRAHE O.F.
ÆDIFICII ET INSTRUMENTORUM
ASTRONOMICORUM STRUCTORIS.
Aº DOMINI 1587, ÆTATIS SUÆ 40.

EFFIGIES TYCHONIS BRAHE O.F.
ÆDIFICII ET INSTRUMENTORUM
ASTRONOMICORUM STRUCTORIS.
Aº DOMINI 1587, ÆTATIS SUÆ 40.

In Prague Tycho became imperial mathematician to Rudolph II; his assistant, Johannes Kepler, carried on his work after his death in 1601. Kepler, who never concealed his admiration for Tycho's observational prowess, once commented that he 'thought in seconds of arc'. Tycho died before Galileo made his telescope, so his instruments had no lenses. To ensure accurate sightings, his measuring instruments had to be very large. These massive instruments – one of them had a radius of about six metres! – were stationary and mounted in the plane of the meridian in order to measure the altitude of stars as they crossed the meridian. With the invention of the telescope, the same precision could be obtained with smaller, more mobile instruments, such as the increasingly popular quadrant (quarter-circle) and sextant (sixth of a circle).

Altitude sextant (opposite, above left), altazimuth semicircle (opposite, above right), solar quadrant (opposite, below left) and altazimuth quadrant (opposite, below right).

That is, in effect, to determine their difference in
longitude. To observe these eclipses accurately, Picard
brought along three refractors with micrometer
eyepieces, in addition to the instruments he had
used to measure the meridian. He set out for Denmark
in July 1671.

Although all that remained of the observatory
Tycho had left behind in 1597 were its foundations,
Picard managed to reconstruct the coordinates of
his instruments and, with the help of Olaus
Roemer, a young Danish astronomer, accomplish
his stated mission.

Roemer's ability impressed Picard, who brought him
back to Paris. He helped him both to secure a position
as astronomy teacher to the son of the king of France
and to gain admission to the Academy of Sciences.

The astronomers straddle the earth in order to measure the distance to the sun

Until Picard's time, the only known astronomical
distance was from the Earth to the moon, which
Aristarchus had measured in the 3rd century BC.
The shape and relative dimensions of planetary orbits
had been established – at any given moment one could
correlate, say, the distance from the Earth to the sun
and the Earth to Mars – but the distances had not
actually been quantified.

In short, the solar system could be mapped out with
correct proportions, but no one knew what
the scale of the map should be. This gap
could be filled simply by measuring any one
distance and then extrapolating all the others from it.

The first step, therefore, was to calculate the distance
between the Earth and some other planet – the closest
one possible. Now, every fifteen or sixteen years,
Mars approaches the Earth at a distance
equal to roughly one-third the distance from
the Earth to the sun. Such
an event was due to

occur in 1672, and a way had to be found to make the most of it.

To measure the distance between the Earth and Mars at a particular moment, Mars would have to be observed simultaneously from two widely separated points on the Earth's surface; then the angle between the two lines of sight would have to be determined. It was already known that, under optimum conditions, this angle would be much smaller than a minute of arc. Therefore, the measurements would have to be extremely precise and the observation sites very, very far apart.

In 1671 Richer leaves for Cayenne, French Guiana. The results of his two-year expedition exceed expectations

To begin with, Richer was the first to make exact observations in a region near the equator, where the sun is very high in the sky and its apparent motion far less subject to atmosphere-related perturbations. The expedition's main objective – measuring the distance from Earth to Mars – was an unqualified success. The angle between the two lines of sight (Paris–Mars and Cayenne–Mars) was a scant 23". Given the coordinates of Cayenne, the result was, in modern units, about 50,000,000 km. Based on this, the scientists determined the distance between the Earth and the sun to be about

Picard's two principal instruments: a quadrant (either vertical or horizontal) to fix the position of reference stars, and a zenith sector with an arc of just a few degrees but a very long radius to measure stellar positions relative to the local vertical (indicated by a plumb line).

133,300,000 km. (Modern scientists calculate the distance at roughly 149,650,000 km.) With these figures, the measurements of the solar system could be calculated. There was even an unexpected bonus. Richer noticed that the same pendulum swung more

A 1699 view of Cayenne and an artist's depiction of its terrain.

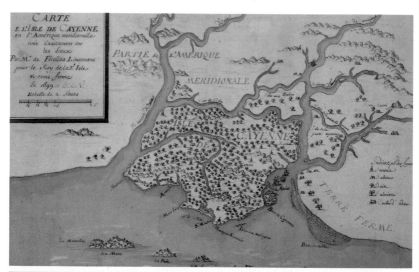

slowly in Cayenne than it did in Paris! This discovery, confirmed in 1682 by an expedition Cassini sent to the Cape Verde Islands (in the Atlantic off western Africa) and the West Indies, was to figure prominently in the debate sparked by Newton's theory of gravitation. Before that, however, Richer's measurements were to have unforeseen repercussions.

OBSERVATIONS ASTRONOMIQUES ET PHYSIQUES

FAITES

EN L'ISLE

DE CAÏENNE.

Now that distance to the sun is known, Roemer is able to measure the speed of light for the first time

Scientists at the time did not know for sure whether light had a finite velocity (then referred to as 'successive motion') or moved through space instantaneously. In any event, even if Galileo was unable to settle the matter, his measurements with lantern signals at least showed that its velocity, if finite, must be very great indeed. Any attempt at calculating it must therefore involve either measuring it at extremely short time intervals or investigating its movement across extremely long distances, that is, astronomical distances.

Richer spent nearly two years in Cayenne. Well equipped and well served by capable assistants, he compiled a substantial amount of data in all areas of interest. The title page of his *Astronomical and Physical Observations* is above.

Now ensconced at the Paris Observatory, Roemer examined the tables of the movements of Jupiter's satellites and noticed recurrent irregularities between their observed and calculated eclipses by the planet. The presumably regular and predictable moments of eclipse occurred earlier than expected six months out of the year (as the Earth moved closer to Jupiter) and later than expected the other six months (as the Earth and Jupiter drew farther apart). The maximum discrepancy any six months apart was about sixteen minutes.

Roemer then realized that the time he recorded for each event coincided not with the moment the eclipse actually took place, but with the moment it was seen on the Earth. If it took light a certain amount of time to cross the distance between Jupiter and the Earth,

then there must be a gap between those two moments, a delay that varies with the distance between the two planets. When the Earth is between the sun and Jupiter, the distance light has to travel between the planets is shorter than when the Earth is beyond the sun – shorter by one diameter of Earth's orbit, a figure Richer had already calculated. A discrepancy of sixteen minutes, or about a thousand seconds, meant that light must travel at a speed of about 300,000 km per second.

Actually, Roemer ended up with a figure closer to 200,000 km per second due to various inaccuracies. The crucial thing, however, was that the order of magnitude of the figure was correct and that light had indeed been found to possess 'successive motion'.

Is it any wonder that Galileo had been unable to solve this problem using lanterns? Over a distance of eleven or twelve km, the time he tried to measure was on the order of only a few hundred-thousandths of a second! In the final analysis, however, Galileo must still be given partial credit for making Roemer's

Born in Aarhus, Denmark, in 1644, Olaus Roemer (above, with his observatory) served as assistant to the professor in charge of publishing Tycho's observation notebooks. Thus, he was the logical choice to assist Picard during the French expedition to the ruins of Uraniborg. Picard brought Roemer back to Paris with him, and the two men spent the next ten years working together, in particular improving a number of instruments. Roemer then returned to his native Denmark, where he perfected several instruments of his own, including a meridian telescope (opposite).

measurement possible. It was he who had discovered those famous satellites of Jupiter and started keeping track of their eclipses.

Roemer published his results in 1675. Scarcely eight years had elapsed since a group of scientists had marked off the meridian at Grand Regard. The astronomers in Paris certainly had made the most of their time.

" At the same time, I should like to inform you that Mr Isaac Newton, Professor of Mathematics at Cambridge, has invented a new kind of telescope. All I can tell you now is that, when first seen and examined here, it was a telescope about six inches long...." And so, in January 1672, Christiaan Huygens learned of Newton's existence in a letter from Henry Oldenburg, secretary of the Royal Society.

CHAPTER 3
FROM THE REFLECTING TELESCOPE TO GRAVITATION

After a civil war, the plague and a fire that destroyed half its old houses, London began to emerge from its ashes in 1667. Newton's telescope (right).

Huygens was by no means Henry Oldenburg's only correspondent at the time – one of the secretary's chief duties was to maintain an uninterrupted flow of letters between the Royal Society and several dozen scientists of all nationalities – but he may have been the most famous.

This medium of official intercourse would soon fall into disuse. In the mid-17th century the number of scientists was fast increasing, and the first scientific journals were starting to appear (*Philosophical Transactions of the Royal Society* in 1664 and *Le Journal des Sçavans* in 1665). In fact, Oldenburg himself served as editor of *Philosophical Transactions* from the first issue to issue 136 (June 1677).

PHILOSOPHICAL
TRANSACTIONS:
GIVING SOME
ACCOMPT
OF THE PRESENT
Undertakings, Studies, and Labours
OF THE
INGENIOUS
IN MANY
CONSIDERABLE PARTS
OF THE
WORLD

Vol I.
For *Anno* 1665, and 1666.

In the *SAVOY*,
Printed by *T. N.* for *John Martyn* at the *Bell*, a little without *Temple-Bar*, and *James Allestry* in *Duck-Lane*, Printers to the *Royal Society*.

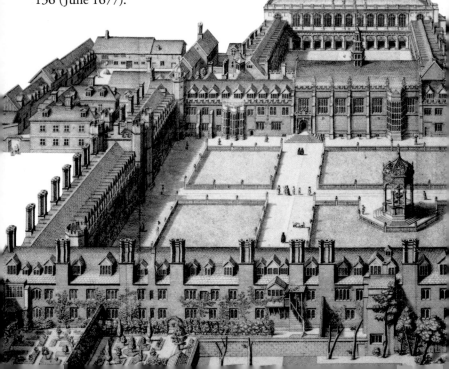

Scientific correspondence and publications alike did more than disseminate information to the general public. They were also means by which to 'ratify' a discovery, to make it official – in a sense, to announce its authorship. Witness the story of the reflecting telescope.

In 1669 an influential professor at Cambridge resigns. He recommends Newton as successor

Newton was only twenty-seven. He had his early work on calculus to thank for this recommendation, as he had yet to tell anyone about his theory of light and colours, much less about gravitation.

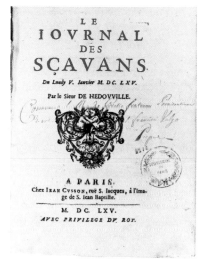

The first scientific journals appeared at about the same time: *Philosophical Transactions* in London in 1664, followed a few months later by *Le Journal des Sçavans* in Paris.

Cambridge University, 1690: a cluster of monasterial buildings attached to what appear to be rows of well-to-do but austere boardinghouses, all surrounded by gardens and lawns that gently slope down to the River Cam. This scene looks much the same today.

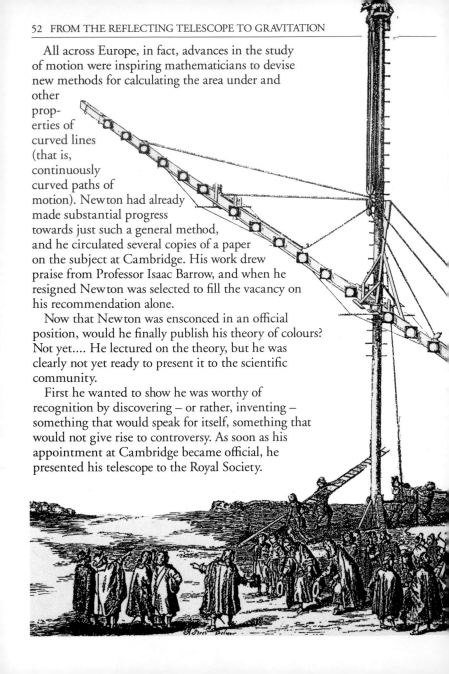

All across Europe, in fact, advances in the study of motion were inspiring mathematicians to devise new methods for calculating the area under and other prop- erties of curved lines (that is, continuously curved paths of motion). Newton had already made substantial progress towards just such a general method, and he circulated several copies of a paper on the subject at Cambridge. His work drew praise from Professor Isaac Barrow, and when he resigned Newton was selected to fill the vacancy on his recommendation alone.

Now that Newton was ensconced in an official position, would he finally publish his theory of colours? Not yet.... He lectured on the theory, but he was clearly not yet ready to present it to the scientific community.

First he wanted to show he was worthy of recognition by discovering – or rather, inventing – something that would speak for itself, something that would not give rise to controversy. As soon as his appointment at Cambridge became official, he presented his telescope to the Royal Society.

Newton builds the first reflecting telescope

Newton had demonstrated that 'white' light is a mixture of light of many colours. The fact that prisms refract the different colours of light by different amounts made this very clear. But the same phenomenon also occurs whenever light rays pass through any piece of glass, particularly lenses; the objective of a refracting telescope always produced images surrounded by coloured fringes (chromatic aberration). As it happened, a way would soon be found to inhibit, if not eliminate, this aberration by combining two lenses made of different kinds of glass. Newton, however, convinced that this could not be done, proposed a radical solution: do away with lenses entirely. Reflectors should replace refractors, he reasoned, because mirrors do not create chromatic aberration.

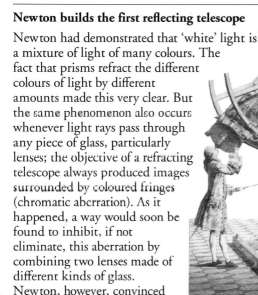

The problem with the reflecting telescope at the time was that it had to be constructed in such a way that the image formed in front of the mirror. To see it, observers would

Although this 18th-century Cassegrain telescope (above) had the same magnifying power as Johannes Hevelius' mammoth refractor (1670, left), it was steadier and easier to handle and it produced brighter images. Magnification was not the sole reason for its gigantic size. The longer a refractor is relative to its diameter, the less pronounced any aberrations will be, particularly coloured images – still an insuperable difficulty in 1670 because neither the reflecting telescope nor achromatic lenses had yet been developed.

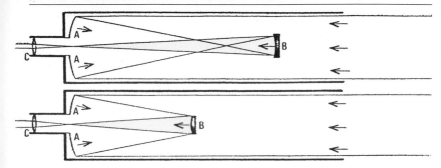

have to position their heads in front of the tube, but that would block the incoming light. Some way had to be found to reflect the image outside the tube, and various different solutions were proposed.

James Gregory (1661) and Guillaume Cassegrain (1672) had suggested that a secondary mirror be placed in the tube to reflect the convergent light back through a hole in the centre of the primary mirror. But Gregory's and Cassegrain's systems called for curved mirrors that could not be manufactured at the time.

Newton came up with the idea of reflecting the convergent light toward the side of the tube instead of straight back. To do this, he placed a flat secondary mirror along the axis of the tube and tilted it at a 45° angle. In this way, the eyepiece would be at the side of the tube, not the end – a practicable solution. Today, large research telescopes are of the Cassegrain variety, but most amateur skygazers, especially those building their own telescopes, use Newtonian reflectors.

Once over this hurdle, Newton still faced others. For one, mirrors were made of a metal alloy, and metal is hard to polish and quick to tarnish. For another, the primary mirror was supposed to be parabolic, but only spherical ones were available then. The discrepancy, albeit small, was enough to produce blurred images (spherical aberration).

Nevertheless, Newton managed to build a little reflecting telescope about twenty cm long. The very compactness of the instrument made it appealing –

In Newton's day, the Gregorian (top) and Cassegrain (bottom) reflecting telescopes existed nowhere but on paper because their small, curved mirrors were too difficult to make. Newton's colleagues were initially won over by the compactness of his instrument. But it was not just a matter of convenience. The smaller the instrument, the more stable it is. A wobbly mounting is so detrimental to image production that a high degree of steadiness was considered even more important than high magnification. Newton's diminutive and elegant instrument could be swivelled in any direction by means of a globe supported on a base and secured by a pair of metal grips.

not the least because the images it produced were nine times bigger than with a refractor four times longer.

In 1671, as soon as the Royal Society published a description of Newton's telescope – in Latin – its inventor became the talk of Europe. Henry Oldenburg, the clearinghouse of Royal Society correspondence,

Images were focused by turning a screw at the lower end of the tube, which moved the objective and the entire back part of the telescope.

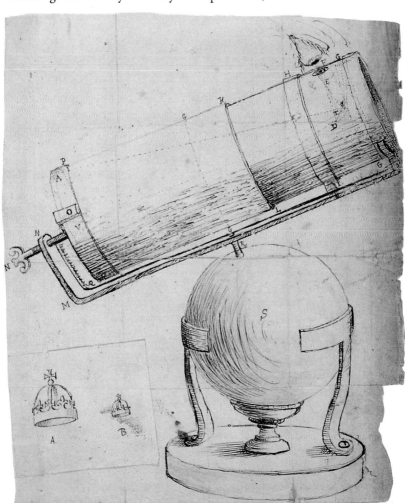

relayed to Newton a steady flow of comments and questions from Christiaan Huygens, Adrien Auzout, John Flamsteed, Johannes Hevelius and James Gregory, and forwarded Newton's replies back to them. Newton became famous overnight. On 11 January 1672 he was elected a fellow of the Royal Society.

In February 1672 Newton finally publishes his theory of light and colours

He did so, of course, in a lengthy letter to Oldenburg. The letter begins with an account of how he purchased a prism and made a hole in a window shutter. It then describes the 'crucial experiment' in detail, argues for the indispensability of reflecting telescopes, and goes on to demonstrate the soundness of his theory by offering a number of experiments that could be replicated at one's leisure.

Oldenburg received this letter the very morning the Royal Society was holding a public meeting. Virtually the entire session was given over to reading and commenting on Newton's letter. The assembly applauded it at length and voted that it be published as soon as three specially selected members repeated the experiments Newton had described.

Two of those three members were Robert Hooke and Robert Boyle, the Royal Society's leading physicists. Hooke had already published an outline of the wave theory of light. Although Newton pointed out in his letter that his colour theory was not based on any particular concept of the nature of light – waves or corpuscles – he intimated that he leaned towards the latter. Consequently, Hooke, while conceding that Newton's findings were of great interest, objected to his interpretation: 'If Mr Newton hath any argument that he supposes an absolute demonstration of his theory, I should be very glad to be convinced of it, the phenomena of light and colours being in my opinion, as well worthy of contemplation as anything else in the world.'

The very thing Newton dreads has happened: opposition

Hooke and Newton would never see eye to eye. Needless to say, Newton hardly saw this as an incentive to divulge his theory of gravitation. He swore that he would not make the same mistake twice and would never publish anything again. He withdrew to his fortress at Trinity College, Cambridge, and became a

Between 1667 and 1672 Newton experimented further with colour to build a stronger case for his theory, which elicited a certain amount of skepticism, to be sure, but no serious opposition.

Gottfried Wilhelm Leibniz (1646–1716) was a German philosopher, theologian, historian and jurist as well as a distinguished mathematician. He and Newton developed differential calculus independently, and Leibniz' method was to become universally adopted.

virtual recluse. But his reputation was already such that his very withdrawal created a stir. What mathematical miracle was Newton silently hatching? People had an inkling because letters to Oldenburg – once again – revealed a nascent dispute with Gottfried Wilhelm Leibniz, the eminent German mathematician. Both of them, it seemed, were independently developing differential calculus, but they were using different methods.

At the same time, Newton was still open to discussing the merits of the reflecting telescope, especially with Huygens. Like Hooke, the Dutch savant was working on a wave theory of light; so that in March, when Oldenburg informed Huygens of Newton's colour theory, Huygens turned a deaf ear and in his reply spoke of nothing but telescopes. Consequently, relations

Greenwich Observatory, central in matters navigational, overlooks the Thames estuary. The prime meridian runs through Greenwich, and Greenwich Mean Time is the world standard.

between the two men remained fairly cordial. In 1673 Huygens sent his treatise on pendulum mechanics to Newton, who thanked him and said that he intended to tell him about his work on curved lines.

Ten years later, Newton's interest was to have repercussions of major importance. It was indirectly responsible for bringing him into contact with the only scientist he felt he could trust, the one who not only finally persuaded him to publish his theory of gravitation, but faithfully supervised its publication himself: Edmund Halley.

Ebullient Halley and reclusive Newton are as different as two people can be

Born in 1656, Halley was fourteen years Newton's junior. His career blossomed even more quickly than

John Flamsteed (1646–1719, above) urged Charles II to found the Royal Observatory, Greenwich (1675), and became its director. He made important contributions to instrumentation and cartography but is chiefly remembered for his catalogue of nearly three thousand stars, the first such compilation in modern astronomy.

the older man's, but Halley's work was limited to matters of astronomy.

The times were conducive to just such a career. Eight years after the Paris Observatory was created, the king of England founded the Greenwich Observatory (in 1675) and appointed as his Astronomer Royal John Flamsteed, who started work on a star catalogue.

Halley's travels and education

An astronomy enthusiast who had the support of a wealthy, cultured father, Halley interrupted his studies at Oxford in 1676 for a two-year stint on St Helena, an island in the Atlantic about 2000 km west of Africa, to catalogue the still little-known stars of the southern hemisphere. It was a useful counterpart to Flamsteed's catalogue. Halley returned to England in 1678, sat his examinations, and was soon elected to the Royal Society. He was only twenty-three.

Despite his youth he was sent to Danzig, Poland, to mediate a dispute between Johannes Hevelius and Robert Hooke.

In 1680 and 1681 he travelled to France and Italy, met Cassini and the other Paris astronomers, and compared his observations of the 1680 comet with theirs – for comets were now streaking across the sky.

The setting of Oxford University, where Halley studied, is as pastoral as ever, and the tower of Magdalen College still rises above the treetops. These days, however, the cows are somewhat farther afield.

Two spectacular comets – one in 1680, one in 1682 – generate excitement among astronomers

Although the comet of 1664 had raised the issue of how the heavens operate, these later sightings gave celestial mechanics fresh urgency. Several comets appear each year, but most are too faint to be seen without sophisticated instruments. After 1618 there were no exceptionally bright comets until 1664 – but it was well worth the wait. Brilliantly visible to the naked eye for six full weeks, it could be detected by even rudimentary instruments for several months. Night after night, astronomers across Europe tracked its course: Auzout in France, Cassini in Rome, Huygens in Holland, Hevelius in Danzig, Hooke in England. Even Newton, then a student, made a note of it.

There was widespread speculation about the paths of comets in particular. Were they circular, as Tycho Brahe had supposed? According to Kepler, they travelled in straight lines: the earth's motion around the sun only made it seem that their paths are curved.

But in 1664 it was clear that the comet's path was actually curved, and Hevelius maintained that it might be elliptical. Furthermore, it was even proposed for the first time that its orbit might be closed – in other words, that the same comet might reappear at regular intervals. The first astronomer to publish this hypothesis was

Continually shifting back and forth between the night sky and a brightly lit piece of paper, star mappers were frustrated by always having to wait for their eyes to adjust to the changing light. The solution was to reduce illumination as much as possible. Not only was the red light from a brazier filled with glowing embers less harsh on the eyes, but it had the added advantage of providing appreciable warmth during early-morning hours.

Pierre Petit of France. In his *Dissertation on the Nature of Comets* (1665), he suggested that the comets of 1618 and 1664 might indeed be one and the same comet.

Hooke seems to have been thinking along the same lines and made his ideas known in March 1665 during a lecture on the comet.

The 1664 comet was a 'retrograde' comet, meaning that its path around the sun ran counter to the motion of the planets and their satellites. In other words, it travelled against the flow of Descartes' 'vortices', raising unavoidable questions about celestial motion and its cause. This problem is mentioned in a notebook from Newton's student days and may have been at least as crucial to his emerging thoughts as the fabled apple.

While we can only speculate about the comet's possible impact on Newton, no such uncertainty surrounds its effect on Hooke and Halley. Witnessing two comets less than twenty years apart put both scientists squarely on the path leading to the theory of universal gravitation.

For Hooke and Halley, the comets of 1664 and 1680, respectively, seem to hint at 'Universal Gravitation'. But it is not the same ...

Although Hooke investigated a wide range of subjects, as did all the

Long thought to be portents of catastrophe in terrifying guises (snakes, flames, fiery swords), comets exemplified the deeply rooted belief that the heavens were a blackboard for divine messages. Because they appeared so suddenly, comets were assumed to be extremely urgent messages that told of imminent disaster. And since hardly a year went by without drought, famine, flood or epidemic, the anticipated calamity seldom failed to materialize.

Im Jahr Christi. 1664. den 24/24 Decemb: in der Nacht gegen Tag, nacht 5. der kleinern Uhr, ward in deß H. Röm Freyen Reichs Stadt Nürnberg, dieser Erschröckliche Comet-Stern wie hier Abgebildet Zuersehen.

scientists of his day, he was primarily interested in the principles of mechanics. The law of physics governing elasticity bears his name. Perhaps that is why he used mechanical devices as models for the attractive force he felt might exist between the bodies of the solar system.

There is a simple apparatus that allows one to describe ellipses – like planetary orbits – arising from attraction towards a centre: it is the familiar pendulum. If a pendulum is pulled away from its resting position and released, it will oscillate, describing a circular arc on either side of the centre of suspension.

If, however, after being pulled away, it is pushed across instead of merely released, it will describe a nearly elliptical curve around the centre and, if the initial impulse is carefully calculated, can even be made to swing in a circle.

The next step: the pendulum as the planetary system?

Hooke investigated this system and exchanged a few letters with Newton on the subject. (Newton's replies were evasive.) Unfortunately, the pendulum turned out to be a not entirely satisfactory model for planetary

An artist's impression of the comet that streaked over Nuremberg on 24 December 1664.

Four comets, as drawn by Johannes Hevelius – brewer, mayor of Danzig and astronomy enthusiast. Hevelius shared his passion for stargazing with his wife. He is credited with several instruments and numerous observations, especially of Saturn and comets.

movement. Its 'centre of attraction'
– its centre of suspension – lies at the
centre of the ellipse, whereas the
orbit of every planet is an ellipse in
which the sun lies at one of its two
foci. Nevertheless, based on this
model, Hooke theorized that
gravitation is inversely proportional
to the distance: thus, a body two
times farther away is attracted with
two times less force.

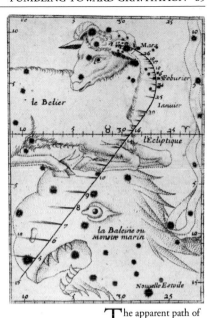

Thinking like an astronomer – and turning to Newton for help

Twenty years later Halley speculated
on the problem of gravitation with
respect to other comets, but as an
astronomer would, starting with
Kepler's third law, relating the size
of a planet's orbit and its period of
revolution: the square of the period
of each planet (the time for completing its journey
around the sun) is proportional to the cube of its mean
distance from the sun.

Assuming these orbits to be circles (which they very
nearly are), Halley showed that this law tallied with the
attraction the sun exerts on the planets, a force
inversely proportional to the square of the distance.

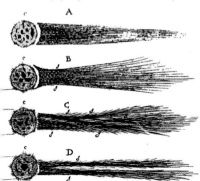

That is, a body
two times farther
away is attracted
with four times less
force, a body three
times farther away
with nine times
less force, and so
on. Halley
wondered if such
an attraction would
be able to account

The apparent path of
a comet, as
illustrated in Pierre Petit's
*Dissertation on the Nature
of Comets* (1665).
Successive observations
made it possible to track
not only the comet's
apparent motion relative
to the constellations, but
the size and direction of
its tail. However, most of
the trajectory pictured
here represents only a few
days' travel, and deducing
the overall configuration
of the comet's orbit from
such a minute segment
was difficult at best.

for Kepler's first law, the one stating that planets describe elliptical orbits.

After attempting – unsuccessfully – to prove this, Halley consulted with Hooke and other colleagues.

The comet of 1680 over Nuremberg: the artist was caught up in the excitement.

No one was able to shed any light on the matter. Finally, in August 1684, he decided to put the question to Isaac Newton, who reputedly knew more about curves and their properties than any man alive.

However brilliant the comet was, it could not have flooded the landscape with light.

In August 1684 Halley went to Cambridge to consult Newton about a problem he was unable to solve and that puzzled the other members of the Royal Society. Newton said he had a solution that he had worked out many years earlier. This and, later on, additional results left no doubt: all motion in the solar system could be explained by a single law, the law of gravitation. Now all Halley had to do was convince Newton to publish it.

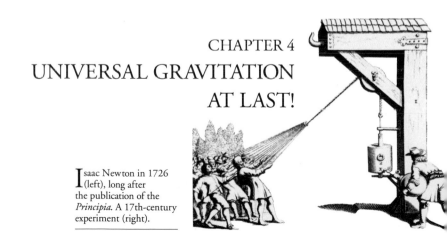

CHAPTER 4
UNIVERSAL GRAVITATION AT LAST!

Isaac Newton in 1726 (left), long after the publication of the *Principia*. A 17th-century experiment (right).

There was nothing on the agenda Newton
had spelled out from 1665 to 1666 that he
had not achieved. He had perfected
methods that expanded calculations once
thought possible only with circles so that
they would be applicable to ellipses.
Coupled with the theory of universal
gravitation, they enabled him to demonstrate
Kepler's laws, until then merely a description of
observed planetary motion. Needless to say, Halley
was dazzled.

With Picard's measurements, Newton can confirm the accuracy of his law of gravitation

Back in 1666, when Newton performed
his calculations involving the apple
and the moon, he had only an
approximate value for the
earth's radius. Picard
accurately calculated
this five years after,
but Newton
did not learn
of it
until
later.

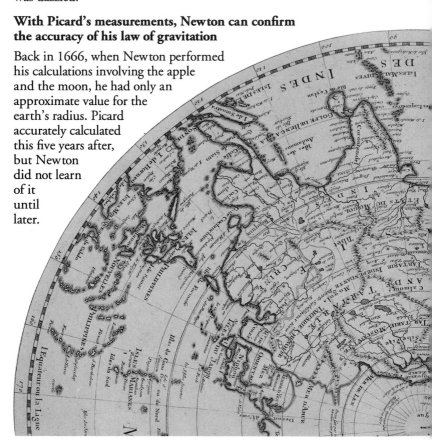

This was so critical, in fact, that it has spawned an anecdote almost as appealing as the story of the falling apple. As soon as he found out about Picard's result, Newton is said to have rushed to his desk to repeat his old calculations on the apple and the moon with the newly determined value for the earth's radius. Everything went so well and the figures led so effortlessly to a confirmation of the law of universal gravitation that Newton became too excited to finish and reportedly asked a friend to complete the calculation for him.

PAR MR. L'ABBE' PICARD.	177
Circonférence de la Terre.	
Toises de Paris.	20541600
Lieuës de 25 au degré.	9000
Lieuës de Marine.	7200
Diametre de la Terre.	
Toises de Paris.	6538594
Lieuës de 25 au degré.	$2864\frac{56}{71}$
Lieuës de Marine.	$2291\frac{59}{71}$

Newton returns to astronomy with the comets of 1680 and 1682

He wrote several letters to Flamsteed, requesting the Astronomer Royal's observation tables of these

It seems odd that Newton did not learn of Picard's results (above) until several years after their publication, because he kept abreast of findings published by French scientists. When Richer observed in Cayenne that a pendulum swung more slowly at the equator – a fact later confirmed (1681–2) by a special expedition the French Academy of Sciences sent to the Cape Verde Islands and the Antilles – Newton found out about it soon afterwards. Members of the expedition were also instructed to measure the longitude of their locations. At the time, there was still no accurate way to do so, and this was a stumbling block to reliable maps.

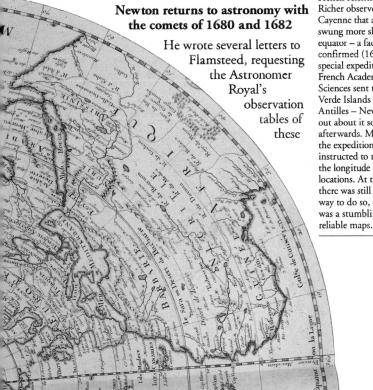

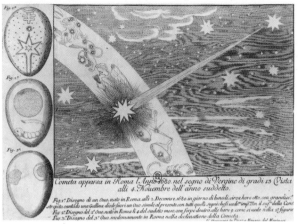

Cometa apparsa in Roma l. Anno 1680. nel segno di Vergine di gradi 13 Vista
alli 4. Nouembre dell' anno suddetto.

Fig: 1° Disegno di un Ouo, nato in Roma alli 5 Decembre 1680 in giorno di lunedi circa hore otto con grandiss.ᵐᵒ
strepito, couduta una Gallina diede fuori un Ouo simile al presente con tutti quelli segni che si uede ͫ ͫⁱᵐⁱˢˢⁱᵐⁱᵐⁱ In il seg: della Comᵉ
Fig: 2° Disegno del 2° Ouo nato in Roma li 4 del suddetto mese con forço dentro alle hore 9 come si uede nella 2ᵃ figura
Fig: 3° Disegno del 3° Ouo medemamente in Roma nella declinatione della Cometa

In Rome extraordinary phenomena reportedly accompanied the appearance of the comet of 1680 in the constellation Virgo. Screeching hens laid eggs that bore mysterious markings thought to be linked to the comet's characteristics and motion. A painstaking record was kept of the date and hour at which each egg was laid. As this engraving clearly shows, belief in the mysterious powers of comets was still very much alive in 1680. Note in particular the plotting of the comet's position with respect to the constellations of the zodiac, then as now a focal point of astrology.

comets as well as the tables Cassini was sending from the Paris Observatory. The curved shape of the path of the 1680 comet was much more pronounced than that of its predecessor in 1664.

Needless to say, there was considerable speculation about the possible causes of this curve. In a series of letters to Newton, Flamsteed argued that the two poles of the sun first attracted, then repelled the comet by magnetic action. Newton countered by pointing out that a 'red-hot lodestone attracts not iron' and doubted that the attraction exerted by the sun on the comet could be magnetic. Furthermore, Newton was unwilling to concede that their interaction might at some stage involve repulsion. He seemed more inclined to subscribe to the theory that the first comet, seen coming and going, may actually have been two different ones.

Comets were at the centre of scientific discussion at the time. In 1682, an even more brilliant comet appeared, and it naturally came under intense scrutiny. By this time, information about the orbit of what was later christened Halley's Comet was precise enough to be used in calculations and eventually provided further confirmation of Newton's thinking.

Cassini observed the comet of 1682 – later christened Halley's Comet – in the gardens of the Paris Observatory. The people depicted here, including several ladies, seem completely engrossed, an indication that comets sparked tremendous excitement even in this scientific setting.

As a rule, Newton keeps his extraordinary discoveries to himself, but Halley induces him to make them public

Of all the reasons for this sudden change of heart, surely the most compelling was the budding friendship between the two men. Perhaps for the first time in his entire life, Newton put his trust in someone – a trust that was never misplaced.

But Newton must also have realized that if he continued to put off publishing his discoveries, someone else might, as it were, jump his claim. Halley was clearly heading towards an inverse-square law of gravitation; he reported that Hooke was, too. Moreover, Newton's rival in mathematics, Leibniz, had just published his work on calculus: a different method, to be sure, but as potentially fruitful as Newton's 'method of fluxions', if not more so, since his was more convenient to use.

Halley returned to London with Newton's pledge that, before the year was out, he would send a copy of the findings he had shown him. They would then be submitted to the Royal Society, not only to record their existence officially but with a view to a far more voluminous work in which Newton would make public all his research in mechanics and its applications to the motion of heavenly bodies.

Even before Halley informs him of the Royal Society's favourable reaction, Newton begins writing the *Principia*, his best-known work

It took Newton less than two years to write the first two books of *Philosophiae Naturalis Principia Mathematica* (or *Mathematical Principles of Natural Philosophy*). (Physics was known as 'natural philosophy' at that time.)

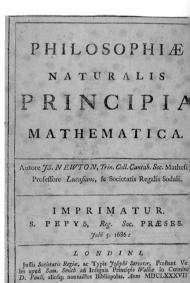

These two books contain the mathematical foundations of his theory of gravitation (in particular, the 'solid sphere' demonstration), as well as the general laws he had formulated to describe motion and how it relates to its causative forces. What we today commonly refer to as Newton's laws of motion were to dominate the entire field of mechanics for more than two hundred years.

Newton had already begun applying the laws to a broad range of phenomena. He investigated collisions, the pendulum, projectiles, air friction, hydrostatics and the propagation of waves (especially sound waves). In short, physics, once inarticulate and disjointed, suddenly blossomed into an organized, consistent, almost architectural whole. The infant Galileo had brought into the world grew up overnight.

Newton had yet to apply his thinking to the motion

of all heavenly bodies – including, of course, comets; he saved that for a third book. But that did not prevent him from working on it while drafting the first two. Late in 1684, he asked Flamsteed for data on the motion of comets as well as information about the orbit size of the satellites of Jupiter and Saturn. It became clear that he had been out of touch with astronomy for years. He was unaware that Cassini had discovered Iapetus in 1671 and Rhea in 1672; for all he knew, Saturn still had only one satellite, Titan, which

The original edition of the *Principia* (1687) with the imprimatur of Samuel Pepys, then president of the Royal Society. As we can see from the sample page below, it made for fairly arduous reading.

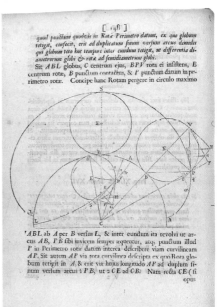

Huygens had discovered in 1656, about the same time he identified Saturn's ring.

Needless to say, the third book (on the movement of the planets and other celestial objects) would be of particular interest to non-mathematicians. Its release was in jeopardy right up to the last, and it took every ounce of Halley's tact to induce Newton to acquiesce to its publication.

A *Principia* manuscript, with corrections in Newton's own hand (centre).

Newton dedicates the *Principia* to the Royal Society, which receives the manuscript of the first two books in April 1686

The society's members resolved at once to publish them. According to English law, any prospective publication required an imprimatur (official authorization) that only a select group of individuals was entitled to grant: the archbishop of Canterbury, the bishop of London, the chancellors of Oxford and Cambridge universities and the president of the Royal Society.

Thus, the society's immediate decision cleared the first obstacle along the road to publication. Now the necessary funds had to be raised, but the society was practically bankrupt at the time. Halley himself shouldered this burden despite personal difficulties arising from his father's death and a protracted battle over his inheritance. To induce Newton to publish, he had probably already pledged to see the book through all stages of publication: discussions with the printer, correction of proofs, checking calculations and diagrams. Now he assumed the financial risk as well. Thus, Halley was, to all intents and purposes, the book's publisher, and one who was especially mindful of the author's wishes. On 7 June 1686 he sent Newton a proof of the first page for the author's comments on paper quality, typography and diagram size. But by that point Newton had shifted his attention to another aspect of universal gravitation: Hooke was claiming he had discovered it first.

Hardly anyone, it

Edmund Halley (1656–1742) brought the *Principia* into the world, but that was only one of many personal accomplishments. He catalogued the southern stars, studied comets, discovered the globular star cluster in Hercules and in 1718 detected the proper motion of stars.

seems, took Hooke seriously – except, of course, a rankled Newton, who dropped the idea of publishing the third book of the *Principia*. He wrote to Halley: '[Physics] is such an impertinently litigious lady, that a man had as good be engaged in lawsuits, as have to do with her. I found it so formerly, and now I am no sooner come near her again, but she gives me warning.'

Needless to say, Halley urged him to reconsider. If Newton relented in the end, no doubt it was largely because he realized that without the third book the *Principia* would be a less salable work and that his continued refusal would more than likely bankrupt his only friend.

In issue 186 of *Philosophical Transactions* (early 1687), Halley, now the journal's editor-in-chief, had the pleasure of writing a rhapsodic review of the *Principia*, adding that it was on sale 'at several booksellers'.

Newton's ideas had finally been put before the public. It remained to be seen what the public would make of them.

The reception of the *Principia*: admired by all, understood by none

Such was the reaction of the 'general public', that is, inquisitive and educated readers who were not professional scientists. The *Principia* is an austere, determinedly mathematical treatise which, except perhaps for the third book, laypeople found virtually unintelligible. It was well received just the same; sales were even brisk. People bravely tried to fathom at least part of it.

This attitude may initially come as a surprise, but it indicates the level of general interest in scientific advances at the time, especially in mathematics. Evidence of this interest crops up everywhere. One case in point is particularly telling. Twenty years before he signed the imprimatur in his official capacity as president of the Royal Society, Samuel Pepys, the noted English diarist, was a secretary

Halley was what we would today call a geophysicist. He produced an exhaustive study of terrestrial magnetism, tides and currents and made significant strides in our understanding of meteorological phenomena. For example, he was the first to explain the trade winds as the vertical motion of air masses caused by varying degrees of solar heating. The first description of the water cycle – evaporation, cloud formation, precipitation, rivers, oceans, evaporation – was also Halley's. Thus, he was the first to suspect the dominant role that heat transfer plays in terrestrial physics.

with the Admiralty. In 1665, his diary tells us, he was taking private lessons in mathematics to learn division!

And this was not solely for his work: he enjoyed it so much, he employed the same teacher to tutor his wife.

This craving was now focused on the *Principia*, which scientists everywhere – with Halley leading the chorus – repeatedly acclaimed as the masterpiece of the new scientific spirit. Since it was a difficult read, the public clamoured for interpreters and popularizers.

Once again, Halley took the lead. In 1687, when King James II was presented with a copy of the first edition of the *Principia*, Halley wrote a long accompanying letter that was immediately published as *The True Theory of the Tides*. In this detailed and readable account of one of the many applications of Newton's theory, he gives the first complete explanation of tidal phenomena as the difference in the moon's attraction on different parts of the earth's surface.

It was a felicitous choice of subject, for it touched upon both physics and astronomy and involved a natural phenomenon with economic ramifications. Without a doubt, Halley was not only a distinguished scientist and shrewd diplomat, but an unsurpassed popularizer! Probably his choice also arose from his love for the sea; he himself later roamed far and wide to chart magnetic variation, study currents and survey ports. But no matter! Love inspires eloquence: dealing as it does with both the *Principia* and the sea, Halley's articulate paper heightened public interest in Newton's book.

Using a secret shorthand that was not deciphered until 1825, Samuel Pepys (1633–1703, left) kept a diary from 1659 to 1669. It is an extraordinary chronicle of daily life in London during an especially eventful period in its history.

Coffeehouses were centres of social and intellectual life in 17th-century London, a time when the moon was the focus of much attention – artistic (below) as well as astronomical.

Scientists and other experts react with the same mixture of respect and incomprehension

In addition to Halley, who obviously knew the *Principia* by heart and had scrutinized it before its publication, in Europe there were perhaps only a few dozen scientists capable of fully understanding it within a reasonable time: Huygens, Leibniz, the unhappy Hooke, and Roemer, among a small elite.

But there were many more who, although unable to follow Newton's demonstrations, could appreciate the significance of his results and overall perspective. However, 'natural philosophy' had to come to grips with an underlying issue, one that was genuinely philosophical in nature.

The whole rationale of Cartesian vortices was an attempt to explain the motion of heavenly bodies as an impulse created by contact with invisible matter. The concept of action at a distance, a force exerted

Immediately after Galileo's death (1642), water pumps became a focal point of scientific debate. Was the underlying principle governing their operation the old Aristotelian concept of *horror vacui?* Or was it air pressure, as Galileo's assistant, Evangelista Torricelli, maintained? In 1648 Blaise Pascal demonstrated the fallacy of the 'nature abhors a vacuum' argument by conducting a series of detailed experiments that were later published in his treatise on hydrostatics.

across millions of kilometres of empty space with no actual contact between the bodies, seemed a throwback to ancient physics, with its 'mysterious properties'. 'Nature abhors a vacuum' (*horror vacui*) was the concept scientists before Evangelista Torricelli, Galileo's assistant, and Blaise Pascal had given to 'explain' the observed rising action of water in pumps.

Newton took exception to this comparison in a letter to Halley. 'It is not I', he wrote, 'who is invoking the *horror vacui*: I simply show that water rises in pumps.' He refused to become embroiled in a dispute on the physical nature of universal gravitation. He had worked out a law of nature in a mathematical form that made it possible to calculate all observed motion. He was very careful not to speculate in the *Principia* (although he did elsewhere) on what the agent that exerts and transmits gravitation might be.

'The most perfect' mechanics, but what about physics?

But his colleagues thought otherwise. They unanimously acclaimed the 'mathematical' structure of Newton's theory but were no less unanimous in their reticence about its philosophical implications, issues which, unlike Newton, they found impossible to sidestep.

The article *Le Journal des Sçavans* published in 1688 to present the *Principia* in France neatly sums up their attitude. While hailing Newton's mechanics as 'the most perfect one can imagine,' the article asked him to 'provide us with a Physics that is as exact as Mechanics,' that is, to declare his views on the true nature of gravitation.

The first tentative efforts in this direction would have to wait until the 20th century and general relativity. Even so, Newton's 'perfect Mechanics' were to triumph time and again in the centuries to come, as scientific discovery succeeded discovery.

This 'jet wagon', propelled by bursts of steam, probably never existed. It is a mid-18th-century demonstration of how Newton's third law – to every action there is an equal and opposite reaction – might be applied. Newton's three laws govern all mechanics, and any apparatus involving motion is as contingent upon those laws now as it was then. While this jet-propelled conveyance is a figment of the imagination, we certainly know very real machines based on the same principle.

All theories must be tested. Explaining an observed phenomenon is one thing, but determining an unexpected phenomenon and then observing it is another. Newton maintained that the earth would be found to be flattened at the poles, and in 1735 the French Academy of Sciences decided to put his prediction to the test.

CHAPTER 5
FROM TRIUMPH TO TRIUMPH

The theory of gravitation enabled scientists to predict the flattening of the earth at the poles, the date of the return of Halley's Comet and the existence of a new planet – all within 150 years!

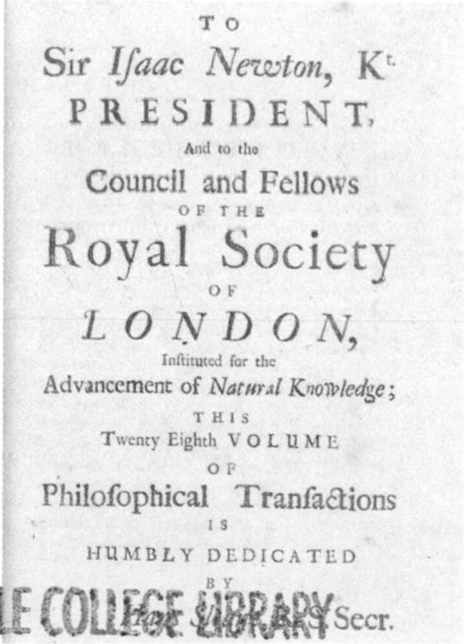

TO
Sir *Ifaac Newton*, K^t
PRESIDENT,
And to the
Council and Fellows
OF THE
Royal Society
OF
LONDON,
Inftituted for the
Advancement of *Natural Knowledge*;
THIS
Twenty Eighth VOLUME
OF
Philofophical Tranfactions
IS
HUMBLY DEDICATED
BY

Despite the inherent contradictions between Newtonian and Cartesian thinking, Newton still enjoyed much respect after the publication of the *Principia* – even in France. In 1699 Newton was elected a foreign associate of the French Academy of Sciences. When he died in 1727, Bernard de Fontenelle, perpetual secretary to the Academy – and Newton's first biographer – delivered his eulogy (see page 122).

The publication of the English scientist's second major work, *Opticks* (1704), had rekindled the long-standing debate between Cartesians and Newtonians, but the differences were far less substantial than is often believed. For example, one of Descartes' closest spiritual heirs, Nicolas Malebranche, took rather a conciliatory stance and promoted scientists who were considerably less 'anti-Newtonian' than one would have expected.

At the time of Newton's funeral, freedom of thought was very

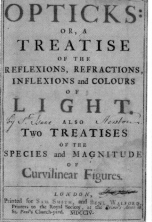

much on the mind of a young Frenchman who had just been released from prison. Banned from Paris at the request of a nobleman who resented a witticism made at his expense, the young man was now an exile in London. His name was François-Marie Voltaire, the author whose *Letters Concerning the English Nation* were to spearhead the Enlightenment. Although Voltaire's admiration for Descartes was undiminished, Newton and his ideas came to represent for him, and for his readers, a struggle between the Ancients and the Moderns, the 'Ancients' being not Descartes, but those who had imprisoned Voltaire when he spoke out.

That is why the expeditions the French Academy sponsored in 1735 and 1736 to test one of Newton's predictions had such far-reaching repercussions. Whether or not the earth is actually flattened at the poles was probably the least of Voltaire's concerns. But to see science triumph over prejudice, to see Newton, an exemplary citizen of the land that then embodied freedom of thought, triumph over the establishment – that mattered to him a great deal.

In 1704 Newton, then sixty-two, published his second major work, entitled simply *Opticks*. It contains a host of theoretical and applied discoveries, as well as an impressive array of experiments of all kinds. Scientists outside England patiently repeated these experiments as soon as the Latin version of the treatise became available.

Richer's and Halley's observations turn the shape of the earth into an issue

Both scientists had noticed that the same pendulum swung

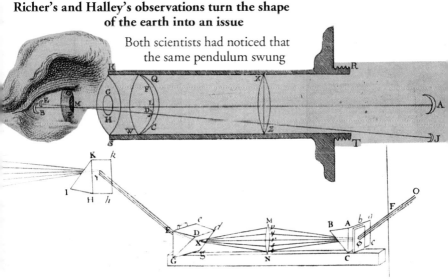

more slowly in the tropics than in Paris or London. During his stay on St Helena, Halley also detected further slowing when he moved the pendulum from the beach to a mountaintop.

If a pendulum swung more slowly near the equator, everyone agreed this phenomenon must also be due to a diminution of gravity. What was causing this effect? Newton demonstrated that a point on the equator is farther away from the earth's centre than Paris or London. That would make the earth not a perfect sphere, but a slightly flattened spheroid, like a sea urchin or a tangerine.

If proven, this 'oblate spheroid' hypothesis would be a decisive argument in favour of Newton's theory. Now, assuming that the earth indeed 'bulged' at the equator, the curvature of a meridian line must be more pronounced at the equator, and the length of a one-degree arc along that meridian shorter. To prove it, the degree of meridian Picard had measured in France would have to be remeasured at the equator. If the arc measured at the equator was shorter, then the earth must be flattened at the poles, and Newton would be proved right.

This was the mission of the French Academy expedition to Peru in 1735.

Since the Peruvian expedition to measure an arc of meridian ran into considerably more obstacles than did its counterpart in Lapland, its work dragged on for more than two years. But the results it obtained were excellent, even better, as it turned out, than those of Maupertuis' team.

Vue de la Base mesurée dans la plaine ...
Sous un arc qui con...
Dessiné du haut de la chu...

NB. On a représenté dans cette vüe tous les objets
compris dans le demi tour de l'horison en supposant
qu'el œil se tournoit successivement vers chacun
d'eux sans sortir du même point.

The Peru expedition runs into delays, and back in Paris Maupertuis grows impatient

Mathematician, biologist, amateur linguist and one of Newton's most ardent supporters in France since 1728, Pierre Louis Moreau de Maupertuis was instrumental in convincing the Academy of Sciences that an expedition should be sent to Peru. Now he succeeded in persuading first the Academy, then the government of Louis XV, to underwrite another expedition, this time north to Lapland (those icy reaches of Norway, Sweden and Finland above the Arctic Circle).

Appointed director of the operation, Maupertuis selected a group of very young assistants to accompany him. For example, Alexis-Claude Clairaut was, at twenty-three, one of the leading mathematicians of his day and already a member of the Academy for five years. Pierre Lemonnier was even younger. The expedition to Lapland had all the hallmarks of a schoolboy's escapade.

In addition to these academicians, the party included a Swedish astronomer,

A portrait of Pierre Louis Moreau de Maupertuis as an Arctic explorer, painted on his return from his expedition.

ni, *depuis* Carabourou *jusqu'à* Oyambaro,
. *degrez de l'horison*,
moulin à foulon d'Yarouqui.

The camp on Mount Niemi, Lapland (left), as drawn by the Abbé Outhier. The 'signal' at the summit is a hollow cone of pine trunks that were stripped of bark so that it could be seen at a distance. This marked one of the vertices of the triangulation system; the others were sighted by means of instruments mounted inside the station, plumb with the tip of the tower. Some trees had to be felled to obtain an unobstructed view in every direction.

Anders Celsius (who acted as interpreter), and the Abbé Réginald Outhier, a cartographer. A congenial soul who was interested in everything, Outhier kept a diary, which he published when he returned. A model of travel literature, it chronicles the activities of the expedition on a day-by-day basis.

The expedition leaves Paris on 20 April 1736 and arrives at Tornea, Finland, in early July

This settlement of no more than a handful of wooden houses marked the southern end of the meridian arc they were about to measure. To the north lay the endless – and uncharted – forests of Lapland. The team's triangulation system would be similar to Picard's, only they would have to locate mountaintops, build reference stations and install sighting instruments under far more gruelling conditions. Fortunately, the Tornea River flowed north to south and could be negotiated by boat – provided someone 'skilful amid cataracts' went along....

Maupertuis hired peasants to serve as guides, rowers, bearers and lumberjacks (whenever trees had to be felled to obtain an unobstructed line of sight from a mountaintop). On 7 July, after the team bought some reindeer skins 'to spread on the earth as our beds', a flotilla of boats laden with valuable instruments and 'such provisions as were thought most necessary' made its way into the river. Six weeks later, despite thick forests, swamps and fearsome Lapp mosquitoes, the triangulation was finished.

At the northern end of the arc lay the riverside hamlet of Kittis. There the team purchased a barn, hauled it to a mountaintop to serve as their final station, and set up a masonry support – Maupertuis

The hamlet of Kittis (opposite, above), at the northern end of the meridian arc chosen for measurement. In this drawing we see various wooden structures that could be found in a northern Scandinavian village at the time, including a tall 'ladder' for drying hay. At the upper left stand buildings that were hauled to the mountaintop for use as an observatory. Since Kittis marked one end of the meridian arc, its astronomical position had to be fixed with the utmost care.

To move about in winter, Maupertuis 'hurtled along at the speed of a half-wild reindeer' in a Lapp sled that often overturned. This sometimes caused the infuriated reindeer to kick; however, as Maupertuis pointed out, 'one can turn the Boat over one, making it serve as a shield against the animal's rage.'

even brought along the cement – for the large
zenith sector, the instrument needed to fix
the astronomical position of Kittis. To do
that, of course, the stars had to be visible;
and the sky refused to clear until the
middle of October. The expedition had
four days before the river froze all the
way to the sea, just enough time for the
return trip to Tornea! The sector was
remounted, again on a masonry support,
to fix Tornea's position. Now rods had to
be made to survey a base line of ten
kilometres along the frozen surface of the river.

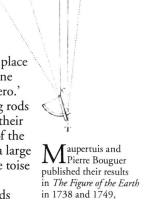

Maupertuis has a Réaumur thermometer and an iron replica of the standard toise

Actually, this iron rod measured a toise only 'in a place
where the thermometer of Monsieur [René Antoine
Ferchault] de Réaumur was at 15 degrees above zero.'
The heating in the room where wooden surveying rods
were carved was regulated accordingly. To adjust their
length to within 'the thickness of a leaf of paper of the
thinnest kind', both ends had to be 'armed with a large
round-headed nail, which was filed away until the toise
exactly fitted the standard.'

On 20 December the entire party set out in sleds
towards the projected base line. It was about minus
20° C, sixty cm of snow blanketed the river ice, and
the sun rose about noon, only to set an hour later.

Nevertheless, the base line was surveyed twice
within a week: the discrepancy between the two
measurements was a scant ten cm.

When they got back to Tornea, all the academicians
had to do was perform the necessary calculations.
(This, as Maupertuis commented, was 'not an operation
difficult in itself'.) It was all over in a few days. In
Lapland an arc of one degree along the meridian
measured 57,395 toises, whereas in France Picard had
come up with 57,060 toises. It seemed the earth was
flattened at the poles after all! Two years later the Peru

Maupertuis and
Pierre Bouguer
published their results
in *The Figure of the Earth*
in 1738 and 1749,
respectively. These books
feature not only the usual
schematic diagrams, but
detailed maps of both
systems of triangulation,
one near Quito, the other
along the Tornea River
in Lapland.

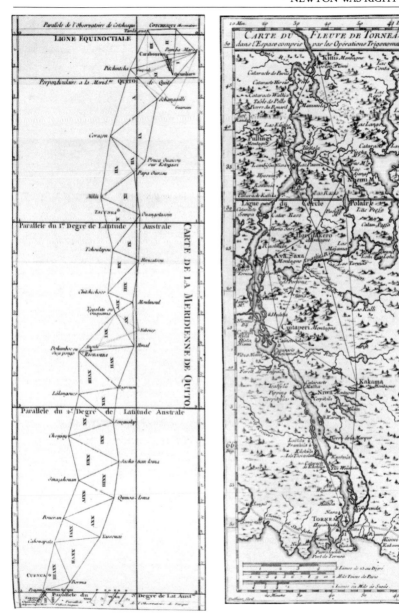

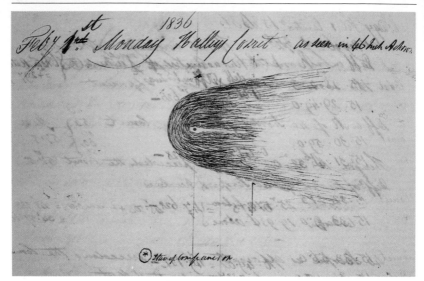

expedition completed its measurements, which corroborated the Lapland results in every respect. Any lingering doubts about the true 'figure of the earth' had been dispelled.

Halley's Comet, as seen from the Greenwich Observatory in 1836.

Five decades after its publication, the *Principia* passes its first test with flying colours. Its second is Halley's Comet

A substantial part of Book III of the *Principia* is devoted to comets. In it Newton maintained that some might have closed paths describing very elongated ellipses and that they could be observed from the earth only when they swung close to the sun. These particular comets should, he argued, always take the same amount of time to complete an orbit and therefore should reappear at regular intervals.

In 1695 Edmund Halley set out to prove this hypothesis by sifting through all recorded observations of comets, at least those recent enough to have had their apparent paths noted. Since the records were based on data from the era of naked-eye observation, his search had to be restricted to the most brilliant comets.

He concluded that the path of the comet of 1682 was very similar to those of the comets of 1607 and 1531. If in fact this was the same comet, it could be expected to return in 1757 or 1758.

In 1705 Halley predicted that the comet would reappear in 'the latter part of 1758 or early in 1759'. People grew more excited as the appointed time drew near: was the world about to witness another confirmation of what were still being referred to as 'Newtonian ideas'? There was some concern among Newtonians on the continent that there might be unavoidable inaccuracies in Halley's calculations.

According to the Bible, a 'star' guided the Magi to the manger. Comet? Nova? The 'star' in this painting, Giotto's *Adoration of the Magi*, is unquestionably Halley's Comet, which appeared in 1301, arousing the usual anxiety and awe.

HAROLD

град̾ цирцагень

въ четвертокъ февраля лѕ дня 1744 года
Изъ циртагены увѣдомляютъ что декабря 28 дня поутру вничилъ въ нихъ вѣто-
лежитъ наньсколько миль отъ помянꙋтаго города въ западъ оно представляла-
ся распространялся а въ востокъ ипроизводилъ тꙋпꙋю ясность что глаза в
шаръ поторо нанецъ поровершеваны повоздухꙋ ꙗвившись вдрꙋгъ раздѣлился на 4 разные
временꙗ въ востокъ съ четверти изападꙋ ипритомъ здѣлался тꙋпꙋ железною громъ
послѣ сего слышны были еще 4 другие также отъ огра понеже сꙗлны намъ первои ичрезъ

The saga of the comets

Comets are not the only unusual sights to manifest themselves in the heavens. Here we see celestial phenomena that were observed in 1743. Two hundred years later, these sightings certainly would have made front-page news! There is more in the sky than stars – the atmosphere, which moves, contains dust and mist, dispersing, diffusing and refracting light and producing haloes, spots, arcs and crosses that are often fleeting but occasionally persistent. The imagination embroiders such events.

Previous pages: This scene from the Bayeux Tapestry shows Halley's Comet in 1066, the year William the Conqueror's expedition fought the Anglo-Saxons at Hastings. It portended disaster – that much was taken for granted – but for which side?

Following pages: The comet in art and in advertising.

DÉCEPTION.

N° 6. — La jolie Comète ne connait plus rien depuis tantôt 75 ans, aussi la Ville de Paris s'empressera de lui montrer les *Grands Magasins du Bon Marché*, dont une visite in...

Therefore they deemed it advisable to repeat the work using the more sophisticated methods that had been developed by the mid-18th century.

Foremost among these Newtonians in France was Alexis-Claude Clairaut, whose reputation, along with his commitment to Newtonian theory, had grown considerably since his return from Lapland. Between 1745 and 1748 he translated the *Principia* into French, with the Marquise du Châtelet, Voltaire's lover and defender. In his annotations to the edition, Clairaut commented that the comet's predicted return in 1758 would be a gratifying moment indeed for Mr Newton's supporters.

PRINCIPES MATHÉMATIQUES
DE LA
PHILOSOPHIE NATURELLE,
Par feue Madame la Marquise DU CHASTELLET.
TOME SECOND.

Small wonder, then, that Clairaut decided to repeat Halley's calculations in 1757, this time factoring in the effects of several planets and using methods of approximation he had spent the past eight years perfecting for other astronomical computations.

Clairaut begins work on what his assistant, Lalande, later calls an 'appallingly lengthy' calculation

They had no computer, but Hortense Lepaute was the next best thing. A calculator for the Paris Observatory, she had spent her entire life drawing up columns of figures. Working full-time, she and Joseph Jérôme Lalande laboured for six months on these calculations.

Hortense Lepaute – or at least her first name – achieved immortality of a more touching kind a few years later. In 1761 a French astronomer named Guillaume Le Gentil left for India to observe the transit of Venus across the disc of the sun, which, using a technique devised by Halley, would enable

" And thus did Truth/To extend her sway/Assume Beauty's countenance/And Eloquence obey. " This quatrain was written in honour of the Marquise du Châtelet, who published a popular treatise on the Newtonian system.

astronomers to measure distances in the solar system. Two such transits, separated by eight years, occur every century.

Le Gentil just missed the transit of 1761 because of delays arising from hostilities between France and England, so he decided to remain in India for the transit of 1769. He built an observatory, learned the language and studied Indian astronomy. In June 1769 the weather there was consistently splendid – except during the transit of Venus, which was obscured by clouds. Le Gentil was exhausted, ailing, driven to despair. He returned to France in 1771 only to learn

In this engraving, Charles Messier (1730–1817) plotted two possible orbits for the return of the comet of 1759. He is best known for his catalogue of celestial objects other than individual stars. The globular star cluster in Hercules is called M13 (Messier 13) and the Andromeda Galaxy, M31 (Messier 31).

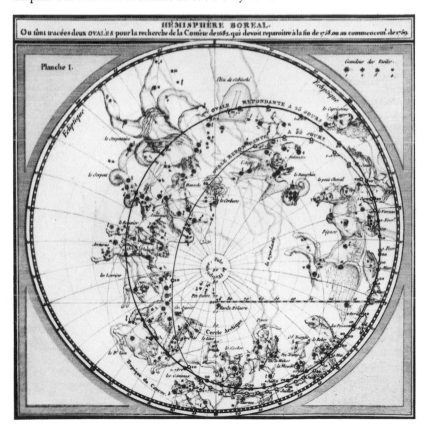

HÉMISPHÈRE BOREAL.

Ou sont tracées deux OVALES pour la recherche de la Comète de 1682 qui devoit reparoître à la fin de 1758 ou au commencem.ᵗ de 1759.

Planche 1.

that he had been declared dead and replaced at the Academy. To make matters worse, the cost of a lawsuit to recover his inheritance nearly bankrupted him.

But he did not return from India empty-handed. He brought back a flower that was unknown in Europe and dedicated it to Hortense Lepaute. The French have called it the *hortensia* (*Hydrangea*) ever since.

Clairaut predicts that the comet will return in mid-April 1759, plus or minus a month

More precisely, he predicted that around that time the comet would reach perihelion, the point of its nearest approach to the sun. And reappear it did, as predicted, reaching perihelion on 14 March, within the margin of error Clairaut had allowed. While his work drew praise from some astronomers, others contended that such elaborate calculations had not been worthwhile; their results were hardly an improvement on Halley's.

In any event, there may have been misgivings in certain quarters about Clairaut's work, but not about Halley's, nor about the fact that the comet's return had offered spectacular proof of Newton's theories. The astronomer Nicolas Louis de Lacaille proposed that it should be named Halley's Comet.

Each time Halley's Comet reappeared, there were improved methods for observing and illustrating the event: engraving in 1835, photography in 1910, computer-enhanced colour images showing temperature differentials in 1986. Its most recent passage was observed at very close range indeed by probes sent up by various nations.

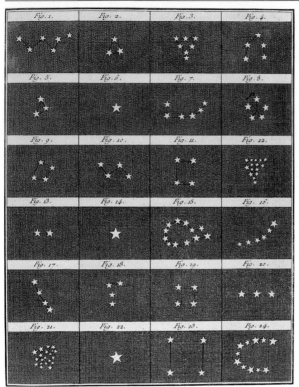

Drawings of the heavens improved, too. When Messier drew the Great Nebula in Orion in 1774 (above), he rendered what he saw in his eyepiece as best he could. It was faint, diffuse, nebulous; so he coined the word 'nebula' to label not just this category of objects (clouds of gas in our galaxy) but all blurred objects, particularly other galaxies (for example, the Andromeda Nebula). By contrast, the later reproduction of ancient drawings of constellations (left) scrupulously gave every star five points.

Since then, each of the comet's punctual reappearances – in 1835, 1910 and 1986 – has reminded us of the man who made the following statement in 1705: 'If the comet should return according to our prediction, about the year 1758, impartial posterity will not refuse to acknowledge that this was first discovered by an Englishman.' Thanks to Lacaille, posterity also remembers that the Englishman in question is Edmund Halley.

Two decades later another Englishman discovers what he thinks is a new comet

If amateur astronomers look to someone as their ideal, if not their patron saint, it is undoubtedly William

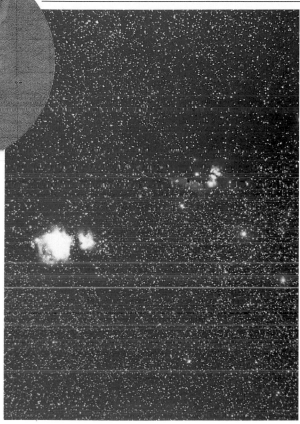

In the late-19th century photography proved a boon to astronomers. How could they have drawn this star field in Orion? Even through his little refractor, Galileo already saw many more stars in the constellation Orion than he could possibly have drawn. Orion does not lie exactly in the Milky Way, but a spiral arm of our own galaxy – hence, millions of faint stars – extends in that direction. This photograph covers a larger field than does Messier's drawing; the Great Nebula is the bright double patch near the left edge. The three stars in Orion's 'belt' are near the right edge (the topmost star is difficult to detect because of another hazy patch). Analysing these tremendous clouds of gas, glowing with the light of the stars within them, is crucial to understanding the evolution of the universe, in particular, the birth of stars.

Herschel. He was a musician by profession – first in a German regimental band, then in the English spa town of Bath, where he played the oboe in an orchestra, served as organist in a chapel and gave music lessons. By day, that is. By night he was an astronomer when skies were clear and a mirror maker when they were overcast: he personally ground and polished the mirrors for his telescopes – more than two hundred of them! People wondered when Herschel ever found time to sleep.

He spent most of his time at night observing stars, with the faithful assistance of his sister Caroline.

Scanning the solar system for as yet undetected objects, he compiled lists of double stars, coloured stars and nebulas. His first major discovery, in 1774, was the Great Nebula in Orion, which modern astrophysicists have since identified as a 'nursery' for new stars.

In the course of his systematic exploration of star fields, he detected a peculiar object. The entry in his log for 13 March 1781 includes a notation about a strange nebulous star that he 'suspected to be a comet'. Since the object moved during the next few nights, Herschel persisted in this belief. On 26 April he submitted his findings to the Royal Society in a paper entitled 'Account of a Comet'.

Astronomers across Europe set to calculating the extremely elongated ellipse the new comet supposedly described, but they were wasting their time. After a few months, there could be no denying the obvious: this

Uranus was a small, diffuse patch in Herschel's telescope (opposite), and even this composite photograph, taken at close range by *Voyager 2*, reveals little in the way of detail. The planet has a dense atmosphere; but its satellites, seen here with extraordinary clarity, do not.

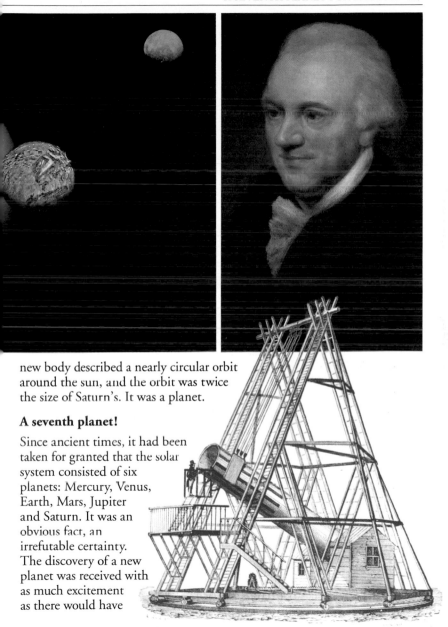

new body described a nearly circular orbit around the sun, and the orbit was twice the size of Saturn's. It was a planet.

A seventh planet!

Since ancient times, it had been taken for granted that the solar system consisted of six planets: Mercury, Venus, Earth, Mars, Jupiter and Saturn. It was an obvious fact, an irrefutable certainty. The discovery of a new planet was received with as much excitement as there would have

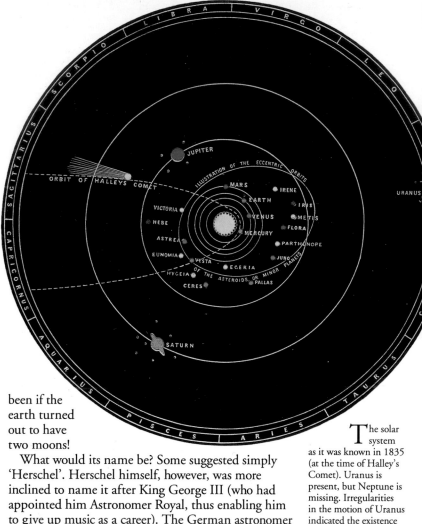

been if the earth turned out to have two moons!

What would its name be? Some suggested simply 'Herschel'. Herschel himself, however, was more inclined to name it after King George III (who had appointed him Astronomer Royal, thus enabling him to give up music as a career). The German astronomer Johann Elert Bode suggested continuing the tradition of naming planets after mythological gods, and his proposal ultimately prevailed. The planet Uranus – Ouranos was 'the sky' in ancient Greek mythology – took its place beyond Jupiter and Saturn.

The solar system as it was known in 1835 (at the time of Halley's Comet). Uranus is present, but Neptune is missing. Irregularities in the motion of Uranus indicated the existence of a more distant planet.

Uranus became the universal topic of discussion all over Europe. For example, a metal discovered a few years later was named uranium in its honour.

However, the planet was not behaving quite the way it should.

Uranus does not quite obey Newton's laws. This eventually leads to the most spectacular find yet!

Irregularities in the motion of Uranus became obvious in 1821, the year that tables for the motion of all the major planets were published. By then, Herschel's planet had not yet completed half its orbit since its discovery (it takes Uranus eighty-four earth years to complete a single revolution around the sun). Curiously enough, however, earlier observations going back to 1690 *were* available! Many astronomers had, in fact, trained their telescopes on a 'star' which, unfortunately for them, they did not attempt to locate at the same spot in the sky the following day. Pierre Charles Lemonnier, for one, had observed Uranus on twelve occasions without ever suspecting that the immortality he so jealously vied with Clairaut to achieve lay at the end of his refractor.

As it turned out, the positional data available in 1821, which involved three successive orbits, did not tally completely with Newton's laws, even factoring in the gravitational pull of the sun and other planets. Therefore, the tables published in 1821 disregarded data prior to Herschel's discovery and based their calculations of its orbit on observations from 1781.

Unfortunately, by 1845 the planet was already 2' of arc off the position the tables had predicted. This removed any lingering doubt: the tables were incorrect. Either Newton's laws were flawed, or Uranus was subject not just to the gravitational pull of the sun and other known bodies in the solar system but to that of a still more distant planet, as yet undiscovered. Would it be possible to apply Newton's laws in reverse and deduce the presence of the 'guilty' planet?

Theoretically, yes. But the necessary calculations were

Urbain Leverrier (1811–77) published his first article on astronomy at the age of twenty. He later studied the motion of Uranus and became the director of the Paris Observatory.

daunting and far more elaborate than even the ones
Clairaut had performed for Halley's Comet.

**In 1845, working independently, John Couch Adams
of England and Urbain Jean Joseph Leverrier of
France start the formidable calculations they hope
will yield a new planet**

The calculations took about a year. Adams had a head
start, came up with a result first, and asked some
English astronomers to see if the unknown planet was
indeed at the place he expected it would be.

Adams had just turned twenty-five and was still a
student. His correspondents had more pressing business
to attend to and put off making the requested
observations.

Leverrier, however, was already a well-known
astronomer. He sent the results of his calculations to
J. G. Galle of the Berlin Observatory, who promptly
trained his refractor on the heavens. That very evening,
he discovered a 'star' that did not appear on a map
that had been released just a few days earlier.

**'Sir, the planet whose position you indicated to me
does indeed exist'**

That was the unprecedented message Leverrier received
late in September 1846.

It was the planet Neptune. Once Leverrier's
discovery was published, people in England realized, to
their dismay, that the unfortunate Adams had predicted
very nearly the same position a few weeks earlier!

In the end, however, it makes little difference to us
whether Leverrier preceded Adams or vice-versa. What
does matter a great deal is that both men had calculated
virtually the same position and that an unknown planet
had been found!

Obviously, by 1845 no one doubted that Newton's
laws governed the workings of the solar system. Even
so, to compute an unknown planet – and then to find
it in the right place!

Almost two centuries after the *annus mirabilis*, the

apple tree at Woolsthorpe had borne its finest fruit. And it would not be its last.

Over the next century new planets were added to the solar system. In 1986 the European Space Agency's *Giotto* came to within a few kilometres of Halley's Comet; in 1989, *Voyager* flew to within a few thousand kilometres of Neptune after successive rendezvous with Jupiter, Saturn and Uranus. Of course, the paths these space probes follow are plotted by computers that are a billion times faster than Hortense Lepaute. But the laws on which their calculations are based – Newton's laws – remain the same.

The motion of an astronaut floating in space, or of *Voyager* hurtling past Saturn, depends solely on Newtonian laws. The earth keeps the astronaut in orbit, while Saturn deflects *Voyager*'s course and speeds it on its way to Uranus.

Overleaf: Halley's Comet, 1986.

DOCUMENTS

TRAITÉ
D'OPTIQUE,
SUR
LA LUMIERE
ET LES COULEURS

LIVRE PREMIER.
PREMIERE PARTIE.

ON dessein dans cet Ouvrage, n'est pas d'expliquer les proprietés de la Lumiere par des Hypotheses; mais de les exposer nuëment pour les prouver par le raisonnement, & par des Experiences. Dans cette vûë je vas commencer par proposer les Definitions & les Axiomes suivants.

A

On the shoulders of giants

At the end of 1675 Newton sent a fairly long letter to the Royal Society in which he described his latest experiments with light, particularly the one involving 'Newton's rings'. At about this time, Robert Hooke suggested a private correspondence.

A letter from Hooke to Newton
These to my much esteemed friend, Mr Isaack Newton, at his chambers in Trinity College in Cambridge

20 January 1676

Sr

The Hearing of a letter of yours read last week in the meeting of ye Royall Society made me suspect yt you might have been some way or other misinformed concerning me and this suspicion was the more prevalent with me, when I called to mind the experience I have formerly had of the like sinister practices. I have therefore taken the freedom wch I hope I may be allowed in philosophicall matters to

Analysing a rainbow, 1747.

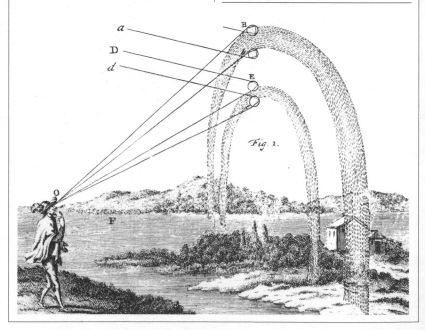

Fig. 1.

acquaint you of myself, first that I do noeways approve of contention or feuding and proving in print, and shall be very unwillingly drawn to such kind of warr. Next that I have a mind very desirous of and very ready to imbrace any truth that shall be discovered though it may much thwart and contradict any opinions or notions I have formerly imbraced as such. Thirdly that I doe justly value your excellent Disquisitions and am extremely well pleased to see those notions promoted and improved which I long since began, but had not time to compleat. That I judge you have gone farther in that affair much than I did, and that as I judge you cannot meet with any subject more worthy your contemplation, so I believe the subject cannot meet with a fitter and more able person to inquire into it than yourself, who are every way accomplished to compleat, rectify and reform what were the sentiments of my younger studies, which I designed to have done somewhat at myself, if my other more troublesome employments would have permitted, though I am sufficiently sensible it would have been with abilities much inferior to yours. Your Designes and myne I suppose aim both at the same thing wch is the Discovery of truth and I suppose we can both endure to hear objections, so as they come not in a manner of open hostility, and have minds equally inclined to yield to the plainest deductions of reason from experiment. If therefore you will please to correspond about such matters by private letter I shall very gladly imbrace it and when I shall have the happiness to peruse your excellent discourse (which I can as yet understand nothing more of by hearing

it cursorily read) I shall if it be not ungrateful to you send you freely my objections, if I have any, or my concurrences, if I am convinced, which is the more likely. This way of contending I believe to be the more philosophicall of the two, for though I confess the collision of two hard-to-yield contenders may produce light yet if they be put together by the ears of other's hands and incentives, it will produce rather ill concomitant heat which serves for no other use but.. . . kindle cole. Sr I hope you will pardon this plainness of your very affectionate humble servt

Robert Hooke

Figures from Newton's *Opticks* (below and on following pages).

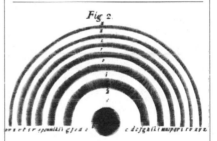

And Newton's response

For his honoured Friend
Mr Robert Hooke
at his Lodgings in Gresham College
London

Cambridge 5 February 1676

Sr

At ye reading of your letter I was exceedingly well pleased & satisfied wth your generous freedom, & think you have done what becomes a true Philosophical spirit. There is nothing

wch I desire to avoyde in matters of
Philosophy more then contention, nor
any kind of contention more then
one in print: & therefore I gladly
embrace your proposal of a private
correspondence. What's done before
many witnesses is seldome wthout
some further concern then that for

formerly tired wth this subject, & have
not yet nor I [believe] ever shall recover
so much love for it as to delight in
spending time about it; yet to have at
once in short ye strongest or most
pertinent Objections that may be made,
I could really desire, & know no man
better able to furnish me wth them

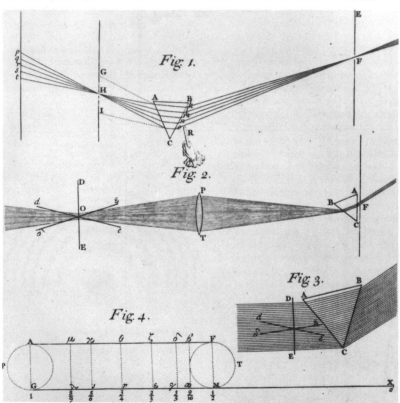

Fig. 1.

Fig. 2.

Fig. 3.

Fig. 4.

truth: but what passes between friends
in private usually deserves ye name of
consultation rather then contest, & so I
hope it will prove between you & me.
Your animadversions will be therefore
very welcome to me: for though I was

then your self. In this you will oblige
me. And if there be any thing els in my
papers in wch you apprehend I have
assumed too much, or not done you
right, if you please to reserve your
sentiments of it for a private letter, I

hope you will find also that I am not so much in love wth philosophical productions but yt I can make them [yield] to equity & friendship. But, in ye meane time you defer too much to my ability for searching into this subject. What Des-Cartes did was a

of two convex glasses & at ye top of a water bubble; and it's probable there may be more, besides others wch I have not made: so yt I have reason to defer as much, or more, in this respect to you as you would do to me, especially considering how much you have been

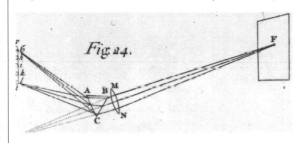

good step. You have added much several ways, & especially in taking ye colours of thin plates into philosophical consideration. If I have seen further it is by standing on ye shoulders of Giants. But I make no question but you have divers very considerable experiments besides those you have published, & some it's very probable the same wth some of those in my late papers. Two at least there are wch I know you have observed, ye dilatation of ye coloured rings by ye obliquation of ye eye, & ye apparition of a black spot at ye contact

diverted by [business]. But not to insist on this: your Letter gives me occasion to inquire concerning an observation you were propounding to me to make here of ye transit of a star near ye Zenith. I came out of London some days sooner then I told you of, it falling out so that I was to meet a friend then at Newmarket, & so missed of your intended directions. Yet I called at your lodgings a day or two before I came away, but missed of you. If therefore you continue in ye mind to have it observed, you may by sending your directions command

Your humble Servant
Is. Newton

Drawings from Newton's *Opticks* showing the refraction of light.

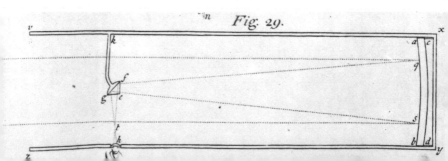

Lapland, day by day

The Abbé Réginald Outhier was chronicler of Maupertuis' expedition as well as its cartographer and illustrator. His Journal of a Voyage to the North in 1736 and 1737 *was published on his return. These excerpts attest to the care and skill that went into the tests they conducted in the spring of 1737 to verify the accuracy of their results.*

Measurement of the arc of meridian.

Easter week: a cartographer's observations

M. de Maupertuis, immediately after his return from Pello, resumed his observations on the lengthening, or diminution of length, of the wooden instruments, from exposure to heat and cold.

During Easter week we observed the variation of the needle, which we found to be 5° and about 5'; it was nearly the same as we noticed in the Baltic before our arrival at Stockholm.

May 24: Checking the direction of the meridian

The sun set entirely at ten minutes past ten. We ascended the highest part of Swentzar: we observed with a quadrant the angle between the sun at the horizon and the signal of Kukama, at the same time counting seconds by a pendulum which we had placed near the spot, in one of those houses used for keeping fodder and cattle in, and which then was empty. The night was very fine: the next morning we returned to take the angle between the rising sun

L app shelter in a pine forest.

on the horizon and the same signal. The direction of our succession of triangles, with respect to the meridian, as found by these observations, differed by some minutes from the direction found at Pello.

We were at first surprised; but quickly reflected that Kittis and Tornea not being under the same meridian, we ought to find some difference, on account of the two meridians approaching sensibly toward the poles in the country where we were. M. Clairaut very quickly made the calculation of what this approximation of the two meridians amounted to; and it was found, by taking this into computation, the directions of the triangles taken at Kittis and Tornea agreed within half a minute of a degree.

Length of a degree of meridian at the Arctic Circle

To make sure the length of the arc of meridian had been measured as accurately as possible, three possible errors had to be obviated: the apparent motion (and proper motion, if any) of the fixed stars; any aberration resulting from the motion of the light radiated by

those stars; and the fact that the 5 1/2° graduated arc of the sector limb was short by 3 3/4". Although Mr Graham had noticed when he graduated the limb that it was too short by that amount and so advised Mr Maupertuis, he 'was desirous of proving it so by experiment' on the third, fourth, fifth and sixth days of May.

On the 4th of May upon the ice of the river we measured a distance of 380 toises [a toise is approximately 2 metres] 1 foot, 3 inches [about 37 cm], which was to serve as a radius. Two firm stakes with two sights . . . were fixed. And having measured the distance between the centres of the two sights, this distance was 36 toises, 6 inches [about 15 cm], 6 2/3 lines, which was to serve as a tangent.

The sextant was placed horizontally in a room upon two firm stocks, supported by an arch . . . And five different observers having taken the angle between the two sights, among whom the greatest difference was not 2", the mean of which being taken, the angle between the two sights was 5°29'48"95. By calculation the angle should have been 5°29'50", that is to say, it differed from the angle observed by 1 1/20".

To take this reading, which was used to check the entire arc of the sector limb, Camus extended a wire along the surface of the graduated limb that marked the divisions. He then extended a second, and by means of these two wires the divisions were observed and tested, one degree at a time.

Abbé Réginald Outhier
Journal of a Voyage to the North in 1736 and 1737, 1744

Richer at Cayenne

In 1672 Jean Richer set up the first modern tropical observatory on the island of Cayenne, French Guiana. In addition to measuring the distance from the Earth to Mars, he compiled a large body of observational data (especially about the sun and Mercury) that were difficult to obtain at higher latitudes. The following excerpt from his observations indicates how far precise astronomical measurement had progressed and reflects Richer's constant concern about synchronizing his clock and the motion of the stars – a concern that led to the auspicious discovery of irregularities in pendulum motion.

Observations from Cayenne

Article V
On Twilight

The duration of twilight at Cayenne is such that I could easily read for forty-five minutes before sunrise and just as long after sunset: this indicates that the refractions of the sun's light are roughly the same here as in France. This is confirmed by the fact that, to see an object clearly with long-distance telescopes, they must be exactly the same length as in Paris.

Not least among my tasks on the island of Cayenne was to observe Mercury, since its motion is not thoroughly known, and because it is seldom visible in Europe and even then only very close to the horizon. I observed this planet on but three occasions; unfortunately, clouds, haze and at other times rain prevented me from doing so more often.

To these observations I shall add such particulars as I was able to record while making them, so as to better determine the location of the planet at the times said observations were made.

The octant was positioned in the meridian in the manner we shall describe later for the observations of 12 September 1672, that is, along a vertical circle 39" from the true meridian and an altitude of 53°44'45".

On 12 September 1672, the west limb of the sun crossed the vertical – very close to the meridian, as we have just mentioned – at 11 h 5°8'28" by the clock, and the east limb at 12 h 0' 36". Consequently, the centre crossed this vertical at 11 h 5°9'32" by the clock; and that same day the meridian altitude of its north limb measured 89°28'15" by octant.

On the evening of the 12th, as Mercury appeared in the West, at the intersection of the two wires in the quadrant eyepiece – which wires meet at right angles, one being vertical, the other horizontal – I measured the altitude of this planet as 15°56'30" when the clock read 6 h 23'15".

While the quadrant remained fixed and steady in this vertical, the Spica of Virgo then crossed the intersection of the above mentioned wires, at which time the altitude of this fixed star was 7°20'0".

I made the following observations in

An early astronomer with an early telescope.

order to show the connection between the daily revolutions of the clock with those of the sun and fixed stars.

On the evening of 12 September 1672, a fixed star marked Baierus E, in the right hand of Aquarius, crossed the meridian at 9 h 2'40" by the clock. The evening of the next day, the 13th, the same fixed star crossed the meridian at 8 h 58'37" by the clock.

On the 14th of the month, with the octant positioned in the same manner as the first observation described above, the west limb of the sun touched the vertical in which the octant lay, vrey close to the meridian at 11 h 57'18" by the clock, and the east limb at 11 h 59'26". Consequently, the centre of the sun crossed the meridian at 11 h 58'22", at which time the meridian altitude of its north limb measured 72°41'10" by octant.

On the evening of the 14th, when the Spica of Virgo crossed the vertical in which the quadrant lay precisely at the point where the vertical and horizontal wires of the eyepiece intersect, its altitude was 10°32'0" at exactly 6 h 46'33" by the clock.

Then Mercury crossed the same vertical, at the same place as the Spica of Virgo, at an altitude of 9°37'10" and at 6 h 47'35" by the clock. The passage of the sun's limbs across the vertical wire of the octant eyepiece – which was, as we have already mentioned, extremely close to the meridian – indicates the time at which this observation was made, as well as the amount by which the movement of the clock needs to be regulated.

Jean Richer
*Astronomical and Physical Observations
from Cayenne*

Fontenelle on Newton: a portrait in words

After Newton's death in 1727 Bernard le Bovier de Fontenelle, secretary of the French Academy of Sciences, delivered the eulogy to which all members were entitled. After a brief account of Newton's formative years and a rather lengthy description of his work (with somewhat greater emphasis on optics than on gravitation), Fontenelle ended his tribute with a physical and psychological portrait of Newton that is a model of its genre.

Isaac Newton's tomb in Westminster Abbey, London.

He was of a middle stature, somewhat inclined to be fat in the latter part of his life; he had a very lively and piercing eye; his countenance was pleasing and venerable at the same time, especially when he pulled off his peruke and shewed his white head of hair that was very thick. He never made use of spectacles, and lost but one tooth in all his life. His name is a sufficient excuse for our giving an account of these minute circumstances.

He was born with a very meek disposition, and an inclination for quietness. He could rather have chosen to have remained in obscurity, than to have the calm of his life disturbed by those storms of Literature, which Wit and Learning brings upon those who set too great a value upon themselves. We find by one of his letters in the *Commercium Epistolicum,* that his treatise of Opticks being ready for the press, certain unseasonable objections which happened to arise made him lay aside this design at that time. 'I upbraided myself' says he, 'with my imprudence, in losing such a reality as Quiet in order to run after a shadow.' But this shadow did not escape him in the conclusion; it did not cost him his quiet which he so much valued, and it proved as much a reality to him as that quiet itself.

A meek disposition naturally promises modesty, and it is affirmed that his was always preserved without any alteration, tho' the whole world conspired against it. He never talked of himself, or with contempt of others, and never gave any reason even to the most malicious observers to suspect him of the least notion of Vanity. In truth he had little need of the trouble and pains of commending himself; but how many others are there who would not have omitted that part, which men so willingly take upon themselves, and do not care to trust with others? How many great men who are universally esteemed, have spoiled the concert of their praise, by mixing their own voices in it!

He had a natural plainness and affability, and always put himself upon a level with every body. Genius's of the first rank never despise those who are beneath them, whilst others contemn even what is above them. He did not think himself dispensed with, either by his merit, or reputation, from any of the ordinary duties of life; he had no singularity either natural or affected, and when it was requisite he knew how to be no more than one of the common rank.

Tho' he was of the Church of England, he was not for persecuting the Non-conformists in order to bring them over to it. He judged of men by their manners, and the true Non-conformists with him were the vicious and the wicked. Not that he relied only on natural religion, for he was persuaded of Revelation; and amongst the various kind of books which he had always in his hands, he read none so constantly as the Bible.

The plenty which he enjoyed, both by his paternal estate, and by his Employments, being still increased by the wise simplicity of his manner of living, gave him opportunities of doing good, which were not neglected. He did not think that giving by his last Will, was indeed giving; so that he left no Will; and he stript himself whenever he performed any act of generosity, either to his Relations or to those whom he thought in want. And the good actions which he did in both capacities were neither few nor inconsiderable. When decency required him upon certain occasions to be expensive and make a shew, he was magnificent with unconcern, and after a very graceful manner. At other times all this pomp, which seems considerable to none but people of a low genius, was laid aside, and the expense reserved for more important occasions. It would really have been a prodigy, for a mind used to reflection and as it were fed with reasoning, to be at the same time fond of this vain magnificence.

He never married, and perhaps he never had leisure to think of it; being immersed in profound and continual studies during the prime of his age, and afterwards engaged in an Employment of great importance, and his intense application never suffered him to be sensible of any void space in his life, or of his having occasion for domestick society.

TOMBEAU A ERIGER A NEWTON

Ⅰᵉʳ PRIX OBTENU EN L'AN MDCCLXXXV A L'ACADEMIE ROYALE D'ARCHITECTURE.

PAR PIERRE JULES DELESPINE.

Tʜese grandiose projects for memorials for Newton, designed in the 18th century by architects Pierre Jules Delespine (above and left) and Etienne-Louis Boullée (below), were never built.

Voltaire on Newton and Descartes

In the fourteenth of his Letters Concerning the English Nation, *published in England in 1732, Voltaire compares Descartes' life and reputation with Newton's: one would be hard put to imagine two more dissimilar men. But Voltaire's admiration for England, where he had found refuge, and for its national hero, Isaac Newton, did not prevent him from giving Descartes his due.*

A Frenchman arriving in London finds things very different, in natural science as in everything else. He has left the world full, he finds it empty. In Paris they see the universe as composed of vortices of subtle matter, in London they see nothing of the kind. For us it is the pressure of the moon that causes the tides of the sea; for the English it is the sea that gravitates towards the moon, so that when you think that the moon should give us a high tide, these gentlemen think you should have a low one. Unfortunately this cannot be verified, for to check this it would have been necessary to examine the moon and the tides at the first moment of creation.

Furthermore, you will note that the sun, which in France doesn't come into the picture at all, here plays its fair share. For your Cartesians everything is moved by an impulsion you don't really understand, for Mr Newton it is by gravitation, the cause of which is hardly better known. In Paris you see the earth shaped like a melon, in London it is flattened on two sides. For a Cartesian light exists in the air, for a Newtonian it comes from the sun in six and a half minutes. Your chemistry performs all its operations with acids, alkalis and subtle matter; gravitation dominates even English chemistry.

The very essence of things has totally changed. You fail to agree both on the definition of the soul and on that of matter. Descartes affirms that the soul is the same thing as thought, and Locke proves to him fairly satisfactorily the opposite.

Descartes also affirms that volume alone makes matter, Newton adds solidity. There you have some appalling clashes.

Non nostrum inter vos tantas componere lites.

This Newton, destroyer of the Cartesian system, died in March last year, 1727. He lived honoured by his compatriots and was buried like a king who had done well by his subjects.

People here have eagerly read and translated into English the eulogy of Newton that M. de Fontenelle delivered in the [French Academy of Sciences]. In England it was expected that the verdict of M. de Fontenelle would be a solemn declaration of the superiority of English natural science. But when it was realized that he compared Descartes with Newton the whole Royal Society in London rose up in arms. Far from agreeing with this judgment they criticized the discourse. Several even (not the most scientific) were shocked by the comparison simply because Descartes was a Frenchman.

It must be admitted that these two great men were very different in their behaviour, their fortune and their philosophy.

Descartes was born with a lively, strong imagination, which made him a remarkable man in his private life as well as in his manner of reasoning. This imagination could not be concealed even in his philosophical writings, where ingenious and brilliant illustrations occur at every moment. Nature had made him almost a poet, and indeed he composed for the Queen of Sweden an entertainment in verse which for the honour of his memory has not been printed.

For a time he tried the career of arms, and having later become the complete philosopher he did not think it unworthy of him to make love. His mistress gave him a daughter named Francine, who died young and whose loss grieved him deeply. Thus he experienced everything pertaining to mankind.

For a long time he believed it necessary to avoid the company of men, and especially his own country, so as to meditate in freedom. He was right; the men of his time did not know enough to enlighten him, and could scarcely do anything but harm him.

He left France because he sought the truth, which was being persecuted there by the wretched philosophy of the School, but he found no more reason in the universities of Holland, to which he retreated. For at the time when in France they condemned the only propositions in his philosophy that were true, he was also persecuted by the self-styled philosophers of Holland, who understood him no better and who, seeing his glory nearer at hand, hated his person the more. He was obliged to leave Utrecht. He was accused of atheism, and this man who had devoted all the penetration of his mind to seeking new proofs of the existence of a God was suspected of not recognizing any.

So much persecution suggested very great merit and a brilliant reputation, and he had both. Reason did pierce a little into the world through the darkness of the School and the prejudices of popular superstition. At length his name became so well known that they tried to lure him back to France by bribes. He was offered a pension of 1000 *écus,* he came on that understanding, paid the fee for a patent which was for sale at that time, did not receive the pension, and returned to work in his solitude in North Holland at the time when the great Galileo, aged eighty, was languishing in the prisons of the Inquisition for having demonstrated the movement of the earth. Finally he died in Stockholm – a premature death caused by faulty diet – amid a few hostile scientists and tended by a doctor who hated him.

The career of Sir Isaac Newton was quite different. He lived to be eighty-five, always tranquil, happy, and honoured in his own country.

His great good fortune was not only to be born in a free country, but at a time when, scholastic extravagances being banished, reason alone was cultivated and society could only be his pupil and not his enemy.

A remarkable contrast between him and Descartes is that in the course of such a long life he had neither passion nor weakness; he never went near any woman. I have heard that confirmed by the doctor and the surgeon who were with him when he died. One can admire Newton for that, but must not blame Descartes.

In England, public opinion of the two of them is that the first was a dreamer and the other a sage.

Very few people in London read Descartes, whose works, practically speaking, have become out of date. Very few read Newton either, because much knowledge is necessary to understand him. However, everybody talks about them, conceding nothing to the Frenchman and everything to the Englishman. There are people who think that if we are no longer content with the abhorrence of a vacuum, if we know that the air has weight, if we use a telescope, it is all due to Newton. Here he is the Hercules of the fable, to whom the ignorant attributed all the deeds of the other heroes.

In a criticism made in London of the discourse of M. de Fontenelle, people have dared to assert that Descartes was not a great mathematician. People who talk like that can be reproached for beating their own nurse. Descartes covered as much ground from the point where he found mathematics to where he took it as Newton after him. He is the first to have found the way of expressing curves by algebraical equations. His mathematics, now common knowledge thanks to him, was in his time so profound that no professor dared undertake to explain it, and only Schooten in Holland and Fermat in France understood it.

He carried this spirit of mathematics and invention into dioptrics, which became in his hands quite a new art, and if he committed some errors it is because a man who discovers new territories cannot suddenly grasp every detail of them: those who come after him and make these lands fertile do at least owe their discovery to him. I will not deny that all the other works of Descartes are full of errors.

Mathematics was a guide that he himself had to some extent formed, and which would certainly have led him in his physical researches, but he finally abandoned this guide and gave himself up to a fixed system. Thereafter his philosophy was nothing more than an ingenious novel, at the best only plausible to ignoramuses. He was wrong about the nature of the soul, proofs of the existence of God, matter, the laws of dynamics, the nature of light; he accepted innate ideas, invented new elements, created a world and made man to his own specification, and it is said, rightly, that Descartes' man is only Descartes' man and far removed from true man.

He carried his metaphysical errors to the point of maintaining that two and two only make four because God has willed it so. But it is not too much to say that he was admirable even in his errors. He was wrong, but at least methodically and with a logical mind; he destroyed the absurd fancies with which youth had been beguiled for two thousand years, he taught the men of his time to reason and to use his own weapons against himself. He did not pay in good money, but it is no small thing to have denounced the counterfeit.

I do not think we really dare compare in any way his philosophy with that of Newton: the first is a sketch, the second is a masterpiece. But the man who set us on the road to the truth is perhaps as noteworthy as the one who since then has been to the end of the road.

Descartes gave sight to the blind; they saw the shortcomings of antiquity and his own. The path he opened has since become measureless. The little book . . . was for a time a complete manual of physics; today all the collected writings of the Academies of Europe put together don't even make a beginning of a system. As one has gone deeper into this abyss it has revealed its infinity. We are about to see what Newton has quarried out of this chasm.

François-Marie Voltaire, 1732

Voltaire.

At home with Herschel

Barthelemy Faujas de Saint-Fond — naturalist, geologist and energetic promoter of hot-air balloons — visited Sir William Herschel in 1784, three years after the Englishman had discovered Uranus. In addition to providing us with a firsthand description of the distinguished astronomer and his estimable sister, his diary records the impressions of a French tourist in London five years before the French Revolution.

On Friday, 13 August [1784], I spent nearly the entire morning writing, or arranging some geological specimens that were given to me. At 1 o'clock I climbed into a carriage with Count Andreani and Mr Thornton for a ride to the Royal Observatory, which is eight miles by carriage from where we are staying, in Howard Street, and the trip cost nineteen pounds.

We found quite a few members of the Royal Society at the Observatory, which they were visiting by royal commission; for in London astronomy is held in the very highest esteem . . . .

This Observatory enjoys the finest of locations, being situated on a lofty hill overlooking the Thames and London. The multitude of craft with which the river is almost entirely covered; the masts mingling with the church spires; the three great bridges across the Thames; the view of St Paul's and a

great many steeples and buildings of every description, make for a sight that is both delightful and astonishing.

The Buildings of the Observatory are plain, unpretentious, devoid of architectural detail and made of brick; but in terms of size, precision and variety the instruments therein leave nothing to be desired. All of these instruments are the largest of their kind.

Mr Maskline was kind enough to show us around and went to great lengths to explain everything to us. Mr Aubert and Mr Sancks introduced us to Mr Herschel, famous for his telescopes and astronomical discoveries, and who was a participant in the commissioned visit to the Observatory. I could not have been more pleased with Mr Herschel, who is as cordial as he is knowledgeable, and who was so kind as to allow me to pay him a visit at his observatory and spend Sunday

The Royal Observatory from Crooms Hill, c. 1680.

night observing the sky with him.

We went out to dine at 4 o'clock at a well-known restaurant where we had a very hearty English-style dinner. Mr Cavendish and Mr Blagden sat next to me, the meal was quite amusing, and we did not leave the table until 7 o'clock, after which we retired to a room where coffee and tea awaited us. The coffee was execrable. Mr Maskline said grace before dinner and again before getting up from the table. The two prayers did not last a minute. I was told this was customary at public meals.

Saturday, 14 August. Went to see a hot-air balloon that had been assembled under the supervision of Mr Schelden, a capable anatomist, and Major Gardinner. This balloon was made of canvas with a glazed finish somewhat akin to oilcloth, of which at first I disapproved on the samples I was shown but found less objectionable when I saw the balloon, which was 56 feet in diameter and spherical in shape. It filled satisfactorily during trial runs, but based on remarks I made it was decided that it should be enlarged; it is due to be inflated to 80 feet and the test flight is set for Friday.

William Herschel.

Mr Herschel's Observatory is in a country house 20 miles from London. I arrived there with Count Andreani and Mr Thornton at 10 o'clock at night.

We found Mr Herschel busy observing in his garden; his sister was in a drawing room perusing Flamsteed's atlas and recording his observations; beside her was a needle-dial pendulum clock that communicated with her brother's telescope by means of a piece of string. This application of fraternal cooperation to the abstract sciences, the meticulous attention of them both, their vitality, their perseverance in their work, the nights given over to observation – these are among the few ideals I shall forever consider myself fortunate to have witnessed.

Mr Herschel's Observatory is by no means above his charming country house: instead he built it himself on a steadier foundation so that his superb instruments would not be subject to the slightest movement. The Observatory is in a garden. Here stands the never-to-be-forgotten telescope with which the planet was discovered, the one Mr Herschel named after the king of England, but which scientists all over Europe unanimously renamed the immortalized Herschel 2. The telescope with which I had the pleasure of making observations for two hours, and saw coloured stars, is 7 feet long and 6 1/2 inches in diameter. Mr Herschel told me that he had personally ground and polished 200 mirrors before bringing it to perfection.

In addition to this telescope, there is another 10 feet long and two others 20 feet long, one of which is 18 3/4 English inches in diameter. The mirror in the latter weighs 150 pounds. This huge machine is mounted on a contrivance so simple and effortless to handle that a child can move it with remarkable ease. This open-air observatory is the most surprising thing imaginable. Whenever Mr Herschel searches for, say, a nebula, or a star of

Eighteenth-century astronomers at work (opposite).

the highest magnitude, he calls from the garden to his sister, who comes to the window straightaway and, consulting one of the large handwritten tables, calls back from the window, 'Near the gamma star', or 'Toward Orion', or some other constellation. In truth, nothing could be more touching and agreeable than this rapport, this straightforward method.

Weightlessness: a perpetual fall

Astronauts 'floating' inside the space shuttle — or in the void beyond — do not lack weight, as the word 'weightlessness' might lead one to believe. They weigh very nearly what they do on earth; that is, the earth attracts them about as strongly as it does an ordinary pedestrian. Since there is nothing to support them, they go into a free-fall, only not like Newton's apple, but like the moon. Their horizontal velocity just offsets the earth's gravitational pull, causing them to swerve inwards. The resulting motion describes a circle that holds them at a constant distance from the earth's centre — and in a perpetual fall.

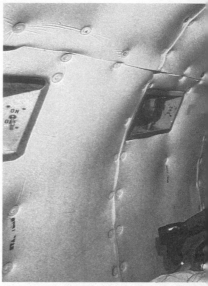

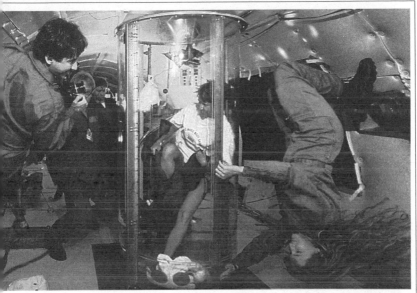

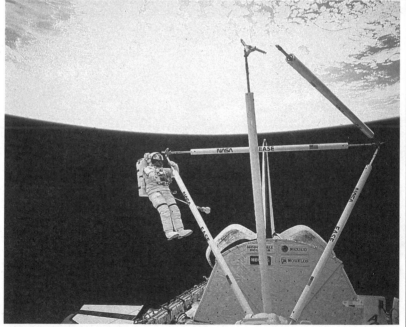

FURTHER READING

Bruno, Leonard, *Landmarks of Science*, 1989

Christianson, Gale E., *In the Presence of the Creator: Isaac Newton and His Times*, 1984

Cohen, I. Bernard, *An Album of Science from Leonardo to Lavoisier*, 1980

Cotardière, Phillippe de la, *Astronomy*, 1987

Fauvel, John, et al., eds., *Let Newton Be!*, 1990

Gjertsen, Derek, *The Newton Handbook*, 1988

Hall, Alfred Rupert, *From Galileo to Newton: 1630–1720*, 1963

Jeans, Sir James, *The Growth of Physical Science*, 1951

More, Louis Trenchard, *Isaac Newton: A Biography*, 1962

Pannekoek, Anton, *A History of Astronomy*, 1989

Pinkerton, John, *A General Collection of the Best and Most Interesting Voyages and Travels in All Parts of the World*, 1808-14

Tancock, Leonard, trans., *Voltaire: Letters on England*, 1980

Turnbull, H. W., ed., *The Correspondence of Isaac Newton*, 1959

Van Helden, Albert, *Measuring the Universe: Cosmic Dimensions from Aristarchus to Halley*, 1986

Westfall, Richard S., *Never at Rest: A Biography of Isaac Newton*, 1981

LIST OF ILLUSTRATIONS

The following abbreviations have been used: *a* above; *b* below; *c* centre; *l* left; *r* right; BCNAM= Bibliothèque du Conservatoire National des Arts et Métiers, Paris; BN=Bibliothèque Nationale, Paris; BO=Bibliothèque de l'Observatoire, Paris

COVER

Front cover Portrait of Newton. Lithograph
Back Pelagio Pelagi. Newton observing the refraction of light. Painting, early 19th century

OPENING

1–7 C. Donato, *Astronomical Observations*. Painting. Pinacoteca, Vatican

CHAPTER 1

10 Newton's house at Woolsthorpe, England. Watercolour, 1840s. Royal Society, London
11 Experiment with light and a prism. Engraving in John Theophilus Desaguliers, *Mathematical Elements of Natural Philosophy*, London, 1747
12 Anonymous. London in 1665 during the Great Plague. Engraving, c. 1665. Magdalene College, Cambridge
13*a* Title page of Kepler's *Harmonices Mundi*, 1609.

BCNAM
13*b* Title page of Galileo's *Dialogus de Systemate Mundi*, 1632. BCNAM
14 Johann Jakob Scheuchzer. Optical effect of a waterfall. Watercolour, c. 1704. Royal Society, London
15*a* Prism. Engraving
15*b* Rainbow observed in Peru. In *Relacion Historica del Viaje a la America Meridional*, Madrid, 1748. Bibliothèque de l'Institut, Paris
16 Goldsmith. Portrait of Newton. Painting, 19th century. Académie des Sciences, Paris
17*a* The spectrum of colours that comprise 'white' light. Engraving in John Theophilus Desaguliers, *Mathematical Elements of Natural Philosophy*, London, 1747
17*b* Experiment with a prism to demonstrate the composite nature of light. *Ibid.*
18 Description of the Universe. Engraving, 1689. Bibliothèque des Arts Décoratifs, Paris
19 Anonymous. Newton and the apple. Chromolithograph, c. 1900
20 Anonymous. Solar system. 18th century. BN
21 Anonymous. Galileo presenting his astronomical telescope to the Doge and the Senate of Venice. Fresco, 1841. La Specola, Florence
22–3 Planisphere of Tycho Brahe. Engraving in *Harmonia Macrocosmi*, 1718
24–5 The Copernican system. *Ibid.*
26 Diagram of Descartes' vortices. Engraving, 17th century. Bibliothèque de Genève
27*a* Moret. Descartes drafting his 'System of the World'. Engraving, 1791. BN

CHAPTER 2

28 Anonymous. *The Paris Observatory in the 17th Century*. Painting. Château de Bussy-Rabutin, Côte-d'Or
29 Henri Testelin. *The Founding of the Académie des Sciences, 1666, and of the Observatory, 1667*. Painting (detail), 1667. Château de Versailles
30–1 Henri Testelin. *Ibid.*
32*a* Christiaan Huygens' pendulum clock. Drawing. BO
32*b* Adrien Auzout's movable-wire screw micrometer. Illustration in *Mémoires de l'Académie des Sciences*. BCNAM
33 B. Vailland. *Christiaan Huygens*. Painting. Hofwyck Museum, Holland
34 A. Coquart. The Paris Observatory. Engraving, 18th century
35*a* A. Coquart. The Paris Observatory and Marly Tower in Gian Domenico Cassini's time. Engraving. BO
35*b* Observing an eclipse. Engraving in Alain Manesson-Mallet, *Description de l'Univers*, 1683. Bibliothèque des Arts Décoratifs, Paris
36–7*b* The Paris Meridian. Engraving. Académie des Sciences, Paris

DOCUMENTS

INDEX

Figures in italic refer to pages on which captions appear.

A

Academy of Sciences, France 30, *30*, 31, 33, 35, 42, *71*, 83, 84, 85, 86, 87, 122
Adams, John Couch 110
Astronomical and Physical Observations (Richer) *45*, 120-1
Auzout, Adrien 29, 31, 32, *32*, 33, 35, 56, 61

B

Barrow, Isaac 52
Bode, Johann Elert 108
Boullée, Etienne-Louis *125*
Boyle, Robert 57
Brahe, Tycho *23*, 38, 39, *39*, *41*, 42, *46*, 61

C

calculus *57*, 58, 73
Cambridge University 11, 12, *51*, 52, 57
Cassegrain, Guillaume *53*, 54, *54*
Cassini, Gian Domenico (Cassini I) *34*, 37, *38*, 45, 60, 61, 72, *73*, 75
Cayenne 43, 44, *44*, *71*, 120–1
Celsius, Anders 89
Châtelet, Marquise du 100, *100*
Clairaut, Alexis-Claude 87, 100, 103, 109, 110
Colbert, Jean Baptiste 30, *30*, 33, 37
comets 26, 61–3, *63*, *64*, 65, *65*, *66*, *67*, 71–2, *72*, *73*, 75, 92, *97*, *98*, *99*; *see also* Halley's Comet
Copernican system *25*
Copernicus, Nicholas 21

D

Delespine, Pierre Jules *125*
Descartes, René 12, 26, *26*, 30, 34, 63, 84, 85, 126–9
Dialogues on the Two Chief World-Systems (Galileo) 12, *13*
Dione 37
Dissertation on the Nature of Comets (Petit) 63, *65*

F

Figure of the Earth, The (Maupertuis and Bouguer) *91*
Flamsteed, John 56, *59*,
60, 71, 72, 75, 132
Fontenelle, Bernard le Bovier de 84, 122–3, 127, 128

G

Galileo 12, *13*, 21, *21*, *26*, 27, 32, *41*, 45, 46–7, 74, *80*, 81, *105*, 127
Galle, J. G. 110
Gassendi, Pierre 30, 35
George III, King of England 108
Great Nebula in Orion *104*, *105*, 106
Greenwich Observatory *58*, *59*, 60, *92*, 130-1 *131*
Gregory, James 54, 56

H

Halley, Edmund 59, 60, *60*, 63, 65, 66, 69, 70, 73, 81, 85, 86, 92, 100, 101; and the *Principia*, 74, 75, 76, 77, 78, 80
Halley's Comet 72, 73, 83, 92–3, *92*, *93*, *97*, 100, *101*, *102*, 103, 104, 110, *111*; *see also* comets
Harmonices mundi (Kepler) *13*
Herschel, Caroline 105, 132-3
Herschel, William *25*, 104–6, 108, 109, 130–3, *132*

Hevelius, Johannes *53*, 56, 60, 61, *64*
Hobbes, Thomas 30
Hooke, Robert 57, 58, 60, 61, 63, 64, 65, 66, 73, 76–7, 80, 114–5
Huygens, Christiaan 20, 31, 32, *32*, *33*, 49, 50, 56, 58, 59, 61, 75, 80

I

Iapetus 37, 75

J

James II, king of England 78
Jupiter 20, 37, *38*, 39, 45, 46, 47, 75, 107, 108, 111

K

Kepler, Johannes 12, *13*, 21, 27, 39, *41*, 61, 65, 66, 70

L

Lacaille, Nicolas Louis de 103, 104
Lapland expedition 87–92, 100, 118–9
law of universal gravitation 18–21, 26–7, 70–1, 74, 76, 81, *83*
laws of motion 74–5, 81
Le Gentil, Guillaume

ACKNOWLEDGMENTS

We are grateful to the Royal Society – especially Keith Moore and Eileen Tweedy – for their cooperation.
Grateful acknowledgment is made for use of material from the following:
Cohen, I. Bernard, ed., *Isaac Newton's Papers and Letters on Natural Philosophy*, reprinted by permission
of the publishers, Harvard University Press, Cambridge, Massachusetts, copyright © 1958, 1978 by the
President and Fellows of Harvard College ('Fontenelle on Newton: a portrait in words'). Letter 14 'On
Descartes and Newton' (pp. 68-72) from *Voltaire: Letters on England*, translated by Leonard Tancock
(Penguin Classics, 1980), copyright © Leonard Tancock, 1980. Reprinted by permission of Penguin Books
Ltd ('Voltaire on Newton and Descartes'). Turnbull, H.W., ed., *The Correspondence of Isaac Newton*, vol.1,
Cambridge University Press, Cambridge, England, 1959 ('On the Shoulders of Giants').

PHOTO CREDITS

Jean-Pierre Maury
was born on 23 September 1937.
A lecturer in physics at the University of Paris–VII,
he has published several books on science,
including, in France,
Galileo: Messenger of the Stars (1986)
and *How the Earth Became Round* (1989).

© Gallimard 1990
English translation © Thames and Hudson Ltd, London,
and Harry N. Abrams, Inc., New York, 1992

Translated by I. Mark Paris

Printed and bound in Italy
by Editoriale Libraria, Trieste